Free Energy Transduction and Biochemical Cycle Kinetics

Terrell L. Hill

DOVER PUBLICATIONS, INC.
Mineola, New York

Bibliographical Note

This Dover edition, first published in 2005, is an unabridged republication of the work originally published by Springer-Verlag New York, Inc., in 1989, which was, in turn, a simplified, abbreviated, and updated version of *Free Energy Transduction in Biology,* published in 1977 by Academic Press, New York.

Library of Congress Cataloging-in-Publication Data

Hill, Terrell L.
  [Free energy transduction in biology]
  Free energy transduction and biochemical cycle kinetics
    p. cm.
  Originally published: Free energy transduction in biology. New York : Academic Press, 1977.
  Includes bibliographical references and index.
  ISBN 0-486-44194-6 (pbk.)
    1. Thermodynamics. 2. Gibbs' free energy. 3. Bioenergetics. 4. Physical biochemistry. I. Title.

QH510.H54 2004
572' .436—dc22

2004059340

Manufactured in the United States of America
Dover Publications, Inc., 31 East 2nd Street, Mineola, N.Y. 11501

Preface

This small book is a simplified, abbreviated, and updated version of the author's *Free Energy Transduction in Biology*, published in 1977 (Academic Press, New York). The present book is meant to be a textbook for a class or for self-study. The first chapter gives a self-contained and elementary discussion of the principles of free energy transduction in biology. Section 5 includes new material on the Onsager coefficients L_{ij} (for systems near equilibrium) not available in 1977. Some readers may wish to study the first chapter only.

The second chapter is a little more sophisticated, and deals with the so-called diagram method for calculating steady-state probabilities and cycle fluxes. Although these concepts are useful in the analysis of free energy transduction systems, they have an intrinsic importance and interest. Section 8 summarizes quite recent new results not included in the 1977 book.

The third chapter is again a step more sophisticated. Some readers may wish to omit it. Free energy levels of the states in a kinetic diagram are introduced. This topic is primarily of conceptual interest for ordinary kinetic diagrams but it is essential in understanding muscle contraction (and related systems) at the molecular level.

Contents

Contents

1

Survey of the Elements of Free Energy Transduction

The primary purpose of this book is to explain the basic principles of free energy transduction in biology in as simple a way as possible. A secondary purpose is to study biochemical kinetic diagrams and cycles. These topics are approached by a consideration of hypothetical model systems, at steady state, that are no more complicated than necessary to bring out the essential points. Also, certain refinements and special topics are intentionally omitted in order to keep the discussion at an elementary level. Thus the limited objective here is to provide the reader with an introductory foundation in these subjects. He or she will then be in a good position to study more sophisticated theory and applications to real systems, as summarized in the more advanced books of Hill[1] (1977), Caplan and Essig[2] (1983), and Westerhoff and van Dam[3] (1987).

A related feature of the present book is that any combination of successive chapters, 1, 12, or 123, provides a coherent story. Chapter 1 contains a survey that uses only very simple mathematics and then the two succeeding chapters add somewhat more detailed quantitative aspects. In fact, Section 1 itself introduces many of the main ideas in a descriptive way.

1. States, Diagrams, Cycles, and Free Energy Transduction

We introduce general concepts here by means of an extended example. Consider a cell that is surrounded by a membrane that separates the cell's interior (inside) from its environment (outside). Suppose that a small molecule M has a much larger concentration inside than outside, $c_{M_i} \gg c_{M_o}$. Another small molecule L has a somewhat larger concentration outside than inside, $c_{L_o} > c_{L_i}$. The inequalities as used here mean, explicitly, $c_{M_i}/c_{M_o} > c_{L_o}/c_{L_i}$. These concentrations are maintained constant (i.e., independent of time). Given a mechanism or pathway across the membrane, M molecules would tend to move spontaneously from inside to outside whereas L molecules would tend to move in the opposite direction. Is it possible for the larger M concentration gradient to be utilized somehow to drive molecules of L from inside to outside *against* the smaller L concentration gradient? If this is accomplished, it is an example of "free energy transduction" (some of the free energy of M is used to increase the free energy of L, as explained in Section 2).

Because of the nature of cell membranes, it is hard to imagine how M and L could perform this trick on their own, using a realistic mechanism. However, this type of activity is commonplace in cells through the mediation of a large protein molecule, or complex of protein molecules, that spans the membrane and interacts suitably with both M and L. Let us denote this large protein molecule or complex by E. The net process accomplished involves M and L only, but the process is made possible only through the intervention or participation of E: E plays the role of an "agent" or "broker" or "middleman."

To illustrate, let us consider a hypothetical but possible model that will accomplish the job specified above. Suppose E exists in two different structures or "conformations," denoted E and E*, which are interconvertible. There is one binding site for L and one for M on both E and E*. However, these sites are accessible only to inside

2

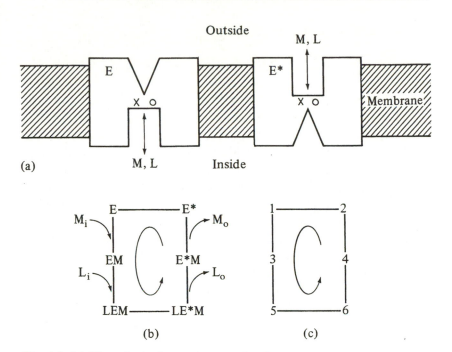

Fig. 1.1. (a) Hypothetical protein complex, in a membrane, with two different conformations, E and E*. × is the binding site for M, o for L. (b) Mechanism for transport of M and L across the membrane. (c) Kinetic diagram associated with the mechanism in (b).

L and M molecules in conformation E, and only to outside L and M molecules in conformation E*. This is shown very schematically in Fig. 1.1(a). Suppose further that L can be bound on its site only if M is already bound on its neighboring site (i.e., the presence of M stabilizes the binding of L). The notation we use for this step is

$$L + EM \rightarrow LEM$$

$$L + E^*M \rightarrow LE^*M.$$

Also, we assume that the binding of L to EM induces the confor-

3

mation change $E \rightarrow E^*$. That is,

$$L + EM \rightarrow LEM \rightarrow LE^*M.$$

L and M are bound less strongly to E^* than to E so they can now be released to the outside.

This model or mechanism is completed and summarized in Fig. 1.1(b). E exists in six states, numbered for later convenience in Fig. 1.1(c). Small molecules are attached to E in some of these states. The lines between pairs of states indicate possible transitions in either direction. The counterclockwise central arrow shows the dominant (but not the only) direction of the transitions.

If one E complex completes one cycle in the counterclockwise direction, the net effect is to transport one M molecule and one L molecule across the membrane from inside to outside. M moves in the direction (downhill) of its concentration gradient, but L is moved against its gradient. Thus, through the participation of E, which is not itself altered by the complete cycle, free energy associated with the concentration gradient of M is used to move L "uphill" in concentration or free energy. Hence the model illustrates transduction of free energy from M to L, or the "active transport" of L, "active" referring to the uphill aspect just mentioned.

With the use of only one cycle in Fig. 1.1(b), there is so-called tight or complete coupling between the movements of M and L. That is, the stoichiometry is exactly one-to-one: each complete cycle moves one M and one L across the membrane.

Figure 1.2(a) is a generalization of the model in Fig. 1.1(b): possible transitions between EM and E^*M are now included. This small modification introduces new complexities. The diagram of states, or "kinetic diagram," Fig. 1.2(a) or 1.2(b), now has three possible cycles, shown in Fig. 1.2(c). The positive direction is arbitrarily chosen as counterclockwise in all three cycles. Because $c_{M_i} > c_{M_o}$, cycle a would operate spontaneously in the positive direction, transporting M from inside to outside. By means of this cycle, E provides a mechanism for the transport of M only. To the extent that cycle a is used, M moves from inside to outside without assisting in the transport of L. From the point of view of free energy transduc-

4

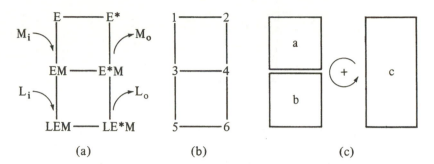

(a) (b) (c)

Fig. 1.2. (a) Generalization of the model in Fig. 1.1(b). (b) Kinetic diagram. (c) Cycles belonging to the diagram.

tion (from M to L), this cycle does not contribute and simply dissipates some of the free energy of M.

Because $c_{L_o} > c_{L_i}$, cycle b would operate spontaneously in the negative (clockwise) direction, moving L from outside to inside. M is not involved (except that it is bound to E in every state of cycle b). Because the object, in this model, is to transport L from inside to outside, clearly cycle b allows movement of L in the wrong direction, and is also a wasteful cycle.

Participation of cycles a and b, as just described, is often referred to as "slippage."

Cycle c in Fig. 1.2(c) is the same cycle as in Fig. 1.1(b). This cycle operates spontaneously in the positive direction and transports L against its concentration gradient at the expense of some of the free energy of M. This is the only cycle in which free energy transduction occurs. Cycles a and b both reduce the efficiency of the free energy transduction. Note also that participation of cycles a and b spoils and exact one-to-one stoichiometry (tight coupling between M and L) that would obtain if cycle c were the only cycle. Thus the transitions EM $\rightleftarrows$ E*M, if they occur, have significant effects.

Let us digress to note Fig. 1.3. This is a considerable generalization of Fig. 1.2(a), with eight states for E and many possible cycles. The heavy cycle is the same as cycle c in Fig. 1.2(c); the arrows show the predominant direction. States in the top square have confor-

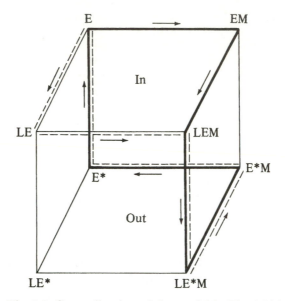

Fig. 1.3. Generalization of the model in Fig. 1.2(a).

mation E and access to the inside; states in the bottom square have conformation E* and access to the outside. The dashed cycle was used as the main cycle in another (more detailed) discussion.[4] In the dashed cycle, L is bound before M, and the subsequent binding of M induces the conformation change E → E*. Of course, the full diagram in Fig. 1.3 allows for the possible participation of *every* cycle in the overall activity of E, interacting with M and L.

Rate Constants and the Stochastic Nature of the Kinetics

We continue to use Fig. 1.2 as an aid in introducing some further important general concepts.

In Figs. 1.2(a) and 1.2(b), there are seven lines in the kinetic diagram and therefore 14 possible transitions. Each transition has a first-order rate constant associated with it, denoted α_{ij} [here we need the numbering of states in Fig. 1.2(b)]. For example, the meaning of

6

$\alpha_{12}dt$ is the following: if a complex is in state 1 (E) at a certain time, the probability that the complex will make the particular transition $1 \to 2$ (E $\to$ E*) in an infinitesimal time interval dt is $\alpha_{12}dt$. Actually, four of the 14 rate constants in Fig. 1.2(a) are "pseudo-first-order." These four are for the transitions $1 \to 3$, $3 \to 5$, $2 \to 4$, and $4 \to 6$. In each of these cases a small molecule is added to E. For example, we have $\alpha_{13} = \alpha_{13}^* c_{M_i}$, where α_{13} is the pseudo-first-order rate constant included among the 14 α_{ij} mentioned above, α_{13}^* is the second-order rate constant for the process E + M (at c_{M_i}) $\to$ EM, and c_{M_i} is the concentration of M on the inside. All concentrations and all the α_{ij} are considered to be constants, independent of time. Typical units employed are ms^{-1} for the α_{ij} and μM^{-1} ms^{-1} or mM^{-1} ms^{-1} for the α_{ij}^*. The concentrations usually have units μM or mM.

Ordinarily we would be interested in a very large ensemble of independent and equivalent E complexes in a cell membrane. If we could select any one of these complexes and follow the detailed succession of states it goes through on the diagram of Fig. 1.2(b), we would observe a random walk from state to state along the lines of the diagram. The complex would spend a random amount of time in a given state (see below) and then jump instantaneously to a neighboring state. Of course, any other E of the ensemble would be doing its own independent random walk, quite oblivious of the activity of its companions. A typical sequence of states in a random walk might be

$$1\ 3\ 5\ 6\ 4\ \underline{3}\ \ 4\ 6\ 4\ 2\ 4\ 2\ \underline{1}\ \ 3\ 1\ 2\ 1\ 3\ 5\ 3\ 4\ 6\ 5\ 6\ 5\ \underline{3}\ \ 1. \qquad (1.1)$$
$$\quad\ \ \ \text{b}+ \qquad\quad\ \text{a}+ \qquad\qquad\qquad\quad\ \ \text{b}-$$

In this particular sequence, three cycles are completed as indicated: b (+ direction), a+, and b−. Cycle b+ moves one L from inside to outside, cycle a+ moves one M from inside to outside, and cycle b− moves one L from outside to inside.

Whenever the random walk reaches, say, state 1, there are two outgoing rate constants, α_{12} and α_{13}. The rate constant for *any* outgoing transition is then $\alpha_{12} + \alpha_{13}$. (States 3 and 4 would have a sum of three outgoing rate constants.) The probability that no transition has occurred after a time t in state 1 is $e^{-(\alpha_{12}+\alpha_{13})t}$. The probability that a transition does occur in the next interval dt is $(\alpha_{12} + \alpha_{13})dt$. Thus, the probability that a transition first occurs between t and $t + dt$ is

$$(\alpha_{12} + \alpha_{13})e^{-(\alpha_{12}+\alpha_{13})t}dt. \tag{1.2}$$

From this, the mean time $\bar{t}_1$ (the subscript refers to state 1) at which a transition first occurs is

$$\bar{t}_1 = \int_0^\infty t(\alpha_{12} + \alpha_{13})e^{-(\alpha_{12}+\alpha_{13})t}dt$$

$$= 1/(\alpha_{12} + \alpha_{13}). \tag{1.3}$$

This is the mean or "expected" time that the random walk spends in state 1, before an outgoing transition, every time state 1 is visited on the walk. When an outgoing transition from state 1 does occur, the probability that it is $1 \rightarrow 2$ is $\alpha_{12}/(\alpha_{12} + \alpha_{13})$ and the probability that it is $1 \rightarrow 3$ is $\alpha_{13}/(\alpha_{12} + \alpha_{13})$. Similar comments can be made about the other five states in the diagram.

If a single complex could be followed in its steady-state random walk over a huge number of transitions (see Section 7), we could observe the fraction of time p_i spent in each state $i = 1, 2, \ldots, 6$. The fraction p_i for state i depends not only on $\bar{t}_i$ but also on the relative frequency that state i is visited during the walk. Also, we could observe the mean frequency or mean rate $J_{a+}, \ldots, J_{c-}$ at which each of the six cycle types $a+, \ldots, c-$ are completed in the course of the walk. These are positive pure numbers per unit time. Typical units for the $J_{\kappa\pm}$ ($\kappa = a, b, c$) are ms^{-1} or s^{-1} (cycle completions are less frequent than transitions). Methods of calculating the p_i and $J_{\kappa\pm}$ from a diagram and its associated α_{ij} will be discussed in Sections 6 and 7.

The net cycle fluxes (+ direction) are defined as $J_\kappa = J_{\kappa+} - J_{\kappa-}$ for any cycle κ. J_κ may be positive or negative.

If we had a large ensemble of E complexes and could observe the particular state $i = 1, 2, \ldots, 6$ of each complex at the same time t, the fraction of complexes in state i would be the same p_i as above.

To be more precise, if time averages (above) for a single E are taken over infinite time and instantaneous ensemble averages are taken over an infinite ensemble, *then* the same p_i would be found by the two methods. The use of finite samples will lead to small differences that can be attributed to fluctuations.

It is interesting that the mean rate of cycle completions ($J_{\kappa \pm}$) cannot be obtained from an instantaneous ensemble average (as above, for the p_i). However, this is not true if an *expanded* diagram, with more details and states, is used (Section 8).

Other Examples of Kinetic Diagrams

We have used Fig. 1.2 to introduce a number of concepts that are actually applicable to a vast array of kinetic systems and diagrams. A few further examples are mentioned here to illustrate this point.

Kinetic diagrams, as in Figs. 1.1, 1.2, and 1.3, are by no means limited to systems involved in free energy transduction or to systems in a membrane. Three extremely simple illustrations of this are shown in Fig. 1.4. Figure 1.4(a) is the diagram for binding a ligand L from solution (concentration c_L) onto a macromolecule E that may be free in solution, or in a membrane, or on a surface. This diagram does not have any cycles. Figure 1.4(b) is the diagram for an enzyme E, possibly free in solution, that binds a substrate S (concentration c_S) and then catalyzes its reaction, S → P, to product (concentration c_P). The dominant direction in the cycle is counterclockwise. The *net* effect of one cycle by one E, in the counterclock-

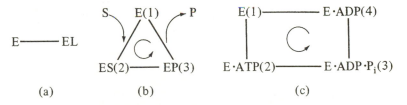

Fig. 1.4. Examples of kinetic diagrams. (a) Binding of L on E. (b) Three-state cycle for conversion of S to P by enzyme E. (c) Four-state cycle for an ATPase, E.

9

wise direction, is the conversion of one S in solution to one P in solution; E is unaffected by a complete cycle. Figure 1.4(c) is an explicit and more elaborate diagram of the same type. Here E is an ATPase, possibly in solution. First ATP is bound to E; then ATP is hydrolyzed on E, to products; then P_i (inorganic phosphate) is released; and finally ADP is released.

In all of these diagrams, transitions can occur in either direction along each line of the diagram, and the transitions are stochastic (random), as governed by the first-order rate constants α_{ij}. Also, characteristically, a macromolecule is the central figure: the diagram enumerates the possible discrete states and transitions of the macromolecule, including small molecules that interact directly with the macromolecule.

Figure 1.5(a) is a modification of Fig. 1.2(a) in which the spontaneous chemical reaction S → P (both species, say, on the inside of the membrane, with concentrations c_S and c_P) is the free energy source and drives L from inside to outside against its concentration

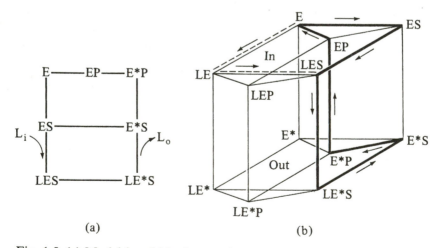

(a)   (b)

Fig. 1.5. (a) Model in which the reaction S → P drives a ligand L across a membrane, against the concentration gradient of L. (b) A more complicated model of the same type that is similar to Fig. 1.3.

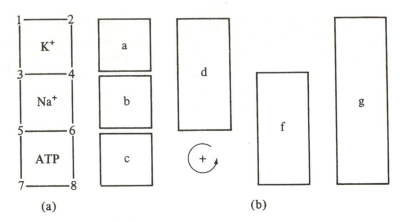

Fig. 1.6. (a) Hypothetical model or diagram for the Na,K-ATPase in a membrane. (b) Cycles belonging to the diagram.

gradient. Slippage may occur because of the possible transitions ES $\rightleftharpoons$ E*S: the reaction S $\rightarrow$ P in the top small cycle accomplishes nothing; the bottom small cycle, running clockwise, allows L to move the wrong way, from outside to inside. Free energy transduction occurs only in the large cycle.

Figure 1.5(b) generalizes Fig. 1.5(a), and is the analogue of Fig. 1.3. The main (large) cycle in Fig. 1.5(a) is marked in Fig. 1.5(b) with heavy lines and arrows. The dashed transitions provide an alternate plausible path (used in ref. 4) between states E and LES.

As a final illustration, we use the semi-schematic and hypothetical model of the Na,K-ATPase complex shown in Fig. 1.6. The diagram has eight states and six cycles; cycle g dominates. Intentionally, the state compositions are not specified explicitly. (Actually, a realistic model would have to be much more complicated than Fig. 1.6.) The complex is in a membrane separating inside from outside (of a cell). For each completed g cycle, in the + (counterclockwise) direction, two K^+ ions are transported out $\rightarrow$ in, three Na^+ ions are transported in $\rightarrow$ out, and one ATP molecule is hydrolyzed. The transitions 3 $\rightleftharpoons$ 4 and 5 $\rightleftharpoons$ 6 introduce some slippage via the other five cycles in Fig. 1.6(b). The slippage spoils the integral 1:2:3 stoichiometry of

11

cycle g. Under normal conditions, K^+ (two per cycle) and Na^+ (three per cycle) would move spontaneously in → out and out → in, respectively, in cycles a, b, and d ($-$ direction), but these ions are driven in the reverse directions by the dominating free energy of ATP hydrolysis in cycles f and g ($+$ direction). Thus this is a classic case of free energy transduction, which occurs only in cycles f and g. Cycle c ($+$ direction) hydrolyzes ATP without accomplishing any transport.

We shall return to this model several times, in greater detail, to illustrate various points.

2. Thermodynamic Forces

The free energies of small molecules have been referred to in a descriptive way in Section 1, in relation to free energy transduction. Here we discuss these free energies more quantitatively. We make use of a succession of examples for this purpose.

Thermodynamic Forces in Fig. 1.2

The term "thermodynamic force" refers to a chemical potential difference that, if operating by itself, determines the direction of a process or processes in a kinetic diagram. In Fig. 1.2, the chemical potentials of M and L on the inside and outside are

$$\mu_{M_i} = \mu_M^o + kT \ln c_{M_i}$$

$$\mu_{M_o} = \mu_M^o + kT \ln c_{M_o}$$

$$\mu_{L_i} = \mu_L^o + kT \ln c_{L_i}$$

$$\mu_{L_o} = \mu_L^o + kT \ln c_{M_o}.$$

(2.1)

These are chemical potentials per molecule (on a per mole basis, replace k by R); μ_M^o and μ_L^o are standard chemical potentials. We are assuming here (and throughout the book), for simplicity, that activity coefficients are approximately unity and that the membrane potential is not involved (M and L are molecules, not ions; or, if they are ions, inside and outside are at the same electrostatic potential). Introduction of membrane potentials is not a problem[1-3] and does not change the general formalities but it *is* a complication, which we avoid. The thermodynamic force driving M from inside to outside is defined as

$$X_M = \mu_{M_i} - \mu_{M_o} = kT \ln(c_{M_i}/c_{M_o}). \qquad (2.2)$$

Similarly, for L, inside to outside:

$$X_L = \mu_{L_i} - \mu_{L_o} = kT \ln(c_{L_i}/c_{L_o}). \qquad (2.3)$$

When X_M is positive, M tends to move spontaneously from inside to outside, etc. At equilibrium, where there is no "drive" in either direction,

$$X_M = 0, \quad \mu_{M_i} = \mu_{M_o}, \quad c_{M_i} = c_{M_o}$$
$$X_L = 0, \quad \mu_{L_i} = \mu_{L_o}, \quad c_{L_i} = c_{I_o}. \qquad (2.4)$$

In the discussion of Fig. 1.2 in Section 1, X_M is positive, X_L is negative, and $X_M > -X_L$.

X_M and X_L are the operational (experimental) forces for this system (Fig. 1.2). Note that X_M and X_L are determined by the concentrations of M and L *in the two solutions*; E is not involved at all in the thermodynamic forces even though E is the central player in the mechanism that allows M and L to cross the membrane.

Each cycle also has a thermodynamic force (some cycles have a zero force). Using the positive cycle direction to define the sign of these forces, we have, in this example,

$$X_a = X_M, \quad X_b = X_L, \quad X_c = X_M + X_L. \qquad (2.5)$$

These relations are determined simply by the individual cycle stoichiometries. In the context of Section 1, X_a and X_c are positive, and X_b is negative. The cycle forces determine the spontaneous directions of the net cycle fluxes (here J_a, $J_c > 0$ and $J_b < 0$).

13

Free energy transduction occurs in cycle c: for each completed cycle (+ direction), the free energy of M drops by an amount X_M; of this, an amount $-X_L$ is used to increase the free energy of L; the remainder,

$$X_M - (-X_L) = X_M + X_L = X_c,$$

is dissipated. The efficiency of free energy transduction, *in cycle c*, is obviously $-X_L/X_M$. Efficiency will be discussed in more detail in Section 4.

At equilibrium, Eq. (2.4), there will be detailed balance in each transition pair (i.e., forward and backward rates are equal):

$$\alpha_{ij} p_i^e = \alpha_{ji} p_j^e. \tag{2.6}$$

The superscript refers to an equilibrium set of state probabilities. If we apply this relation to each transition in cycle a, and multiply, we have

$$\frac{\alpha_{13}\alpha_{34}\alpha_{42}\alpha_{21}}{\alpha_{31}\alpha_{43}\alpha_{24}\alpha_{12}} = \frac{p_3^e p_4^e p_2^e p_1^e}{p_1^e p_3^e p_4^e p_2^e} = 1, \tag{2.7}$$

or

$$\alpha_{13}^* c_{M_i}\alpha_{34}\alpha_{42}\alpha_{21} = \alpha_{31}\alpha_{43}\alpha_{24}^* c_{M_o}\alpha_{12},$$

or

$$\alpha_{13}^*\alpha_{34}\alpha_{42}\alpha_{21} = \alpha_{31}\alpha_{43}\alpha_{24}^*\alpha_{12}. \tag{2.8}$$

The concentrations cancel at equilibrium. Equation (2.8) is an intrinsic relation among the rate constants of cycle a that holds under any conditions, for example, steady state or a transient (even though it was derived using detailed balance). That is, the rate constants in the cycle are not completely independent of each other; in a self-consistent model they are required to satisfy this relationship.

Let us consider cycle a now at an arbitrary nonequilibrium steady

14

state ($c_{M_i} \neq c_{M_o}$). We define

$$\Pi_{a+} = \alpha_{13}\alpha_{34}\alpha_{42}\alpha_{21}, \quad \Pi_{a-} = \alpha_{31}\alpha_{43}\alpha_{24}\alpha_{12}. \tag{2.9}$$

That is, for any cycle κ, $\Pi_{\kappa+}$ is defined as the product of first-order rate constants around cycle κ in the positive direction, etc. Then, using Eqs. (2.2), (2.5), and (2.8),

$$\frac{\Pi_{a+}}{\Pi_{a-}} = \frac{c_{M_i}}{c_{M_o}} = e^{X_M/kT} = e^{X_a/kT}. \tag{2.10}$$

Thus there is a simple relation between the quotient of rate constant products, Π_{a+}/Π_{a-}, and X_a. Analogous consideration of cycles b and c in Fig. 1.2 leads to

$$\frac{\Pi_{b+}}{\Pi_{b-}} = \frac{c_{L_i}}{c_{L_o}} = e^{X_L/kT} = e^{X_b/kT} \tag{2.11}$$

$$\frac{\Pi_{c+}}{\Pi_{c-}} = \frac{c_{M_i}c_{L_i}}{c_{M_o}c_{L_o}} = e^{(X_M+X_L)/kT} = e^{X_c/kT}. \tag{2.12}$$

This type of relationship (between $\Pi_{\kappa+}/\Pi_{\kappa-}$ and X_κ) is found for every kind of cycle (see below). Not surprisingly, there is some thermodynamics embedded in the rate constants.

The Thermodynamic Force in Fig. 1.4(b)

Here there is a chemical reaction S → P catalyzed by E, but no free energy transduction. The concentrations of S and P in solution are c_S and c_P. The thermodynamic force that drives the only cycle (in the positive direction) is also the operational force:

$$X = \mu_S - \mu_P \tag{2.13}$$

$$\mu_S = \mu_S^\circ + kT \ln c_S$$

$$\mu_P = \mu_P^\circ + kT \ln c_P, \tag{2.14}$$

where μ_S and μ_P are the chemical potentials of S and P in solution. Thus,

$$X = \mu_S^o - \mu_P^o + kT \ln(c_S/c_P). \tag{2.15}$$

The enzyme E is not involved in X. At equilibrium,

$$X = 0, \quad \mu_S = \mu_P, \quad K = \frac{c_P}{c_S} = e^{-(\mu_P^o - \mu_S^o)/kT}. \tag{2.16}$$

K is the conventional equilibrium constant for the reaction $S \rightleftarrows P$ in solution and $\mu_P^o - \mu_S^o$ is the associated standard free energy change. If we use the relation between K and $\mu_P^o - \mu_S^o$ in Eq. (2.16), X in Eq. (2.15) can be expressed (for *arbitrary* c_S and c_P) as

$$X = kT \ln(Kc_S/c_P). \tag{2.17}$$

Obviously, $X = 0$ when c_P/c_S is equal to the equilibrium ratio. Equation (2.17) is the analogue of Eqs. (2.2) and (2.3); in these earlier equations, in effect, $K = 1$ [see Eq. (2.4)].

We now apply Eq. (2.6) to the three transition pairs in the cycle, at equilibrium:

$$\frac{\alpha_{12}\alpha_{23}\alpha_{31}}{\alpha_{21}\alpha_{32}\alpha_{13}} = \frac{p_2^e p_3^e p_1^e}{p_1^e p_2^e p_3^e} = 1,$$

or

$$\alpha_{12}^* c_S \alpha_{23}\alpha_{31} = \alpha_{21}\alpha_{32}\alpha_{13}^* c_P,$$

or

$$\alpha_{12}^*\alpha_{23}\alpha_{31} = \alpha_{21}\alpha_{32}\alpha_{13}^* K. \tag{2.18}$$

Equation (2.18) is a required relation among rate constants for this model. This equation can also be used to express K in terms of rate constants. At an arbitrary steady state ($c_P/c_S \neq K$), use of Eq. (2.18) leads to

$$\frac{\Pi_+}{\Pi_-} = \frac{\alpha_{12}\alpha_{23}\alpha_{31}}{\alpha_{21}\alpha_{32}\alpha_{13}} = \frac{\alpha_{12}^* c_S \alpha_{23}\alpha_{31}}{\alpha_{21}\alpha_{32}\alpha_{13}^* c_P} = \frac{Kc_S}{c_P}. \tag{2.19}$$

Thus we find, from Eq. (2.17),

$$\frac{\Pi_+}{\Pi_-} = e^{X/kT}, \tag{2.20}$$

just as in Eqs. (2.11) and (2.12).

The Thermodynamic Force in Fig. 1.4(c)

The ATPase cycle in Fig. 1.4(c), with enzyme E, is a special case of Fig. 1.4(b) in which the product P has two components, ADP and P_i. The substrate S is ATP. We denote the concentrations of ATP, ADP, and P_i in solution by c_T, c_D, and c_P, respectively. The separate chemical potentials are then

$$\mu_T = \mu_T^o + kT \ln c_T, \tag{2.21}$$

etc. The thermodynamic force driving the four-state cycle (+ direction) is

$$X_T = \mu_T - \mu_D - \mu_P$$
$$= \mu_T^o - \mu_D^o - \mu_P^o + kT \ln \frac{c_T}{c_D c_P}. \tag{2.22}$$

The subscript on X (used below) refers to the hydrolysis of ATP. Again, the enzyme is not involved in this force. At equilibrium, the conventional relations for the reaction $T \rightarrow D + P_i$ are

$$X_T = 0, \quad K = \frac{c_D c_P}{c_T} = e^{-(\mu_D^o + \mu_P^o - \mu_T^o)/kT}. \tag{2.23}$$

Hence, with arbitrary steady-state concentrations c_T, c_D, and c_P, Eq. (2.22) can be rewritten in the form

$$X_T = kT \ln \frac{Kc_T}{c_D c_P}. \tag{2.24}$$

This is an obvious extension of Eq.(2.17).

Consideration of detailed balance at equilibrium, as before, yields

$$\frac{\alpha_{12}^* c_T \alpha_{23} \alpha_{34} \alpha_{41}}{\alpha_{21} \alpha_{32} \alpha_{43}^* c_P \alpha_{14}^* c_D} = 1,$$

or

$$\alpha_{12}^* \alpha_{23} \alpha_{34} \alpha_{41} = \alpha_{21} \alpha_{32} \alpha_{43}^* \alpha_{14}^* K. \tag{2.25}$$

Then we find, at an arbitrary steady state,

$$\frac{\Pi_+}{\Pi_-} = \frac{\alpha_{12} \alpha_{23} \alpha_{34} \alpha_{41}}{\alpha_{21} \alpha_{32} \alpha_{43} \alpha_{14}} = \frac{Kc_T}{c_D c_P} = e^{X_T/kT}. \tag{2.26}$$

We see in all of these examples, whether the small-molecule process involves simple transport or a chemical reaction, for any cycle κ,

$$\frac{\Pi_{\kappa+}}{\Pi_{\kappa-}} = e^{X_\kappa/kT}. \tag{2.27}$$

X_κ is the thermodynamic force in cycle κ (+ direction). This force is determined by small molecule concentrations in solution only; E is not involved. However, E obviously plays a vital role in the rate constants α_{ij} and hence in the separate products $\Pi_{\kappa+}$ and $\Pi_{\kappa-}$. Thus the influence of E cancels out on taking the quotient $\Pi_{\kappa+}/\Pi_{\kappa-}$.

Cycle forces are combinations of operational forces, as in Eqs. (2.5). Hence, in constructing a model, the cycle forces are generally known in advance. The first-order rate constants of the model, using the actual small-molecule concentrations at the steady state of interest, must be chosen so that they are consistent with Eq. (2.27) for each cycle in the diagram.

Thermodynamic Forces in Fig. 1.6

A simple Na,K-ATPase model was introduced at the end of Section 1. We return to it here, briefly. The new feature is stoichiometry. Recall that, in cycles a, b, and c, in the positive direction, two K^+ are moved out → in per cycle a, three Na^+ are moved in → out per cycle b, and one ATP is hydrolyzed per cycle c. The respective operational thermodynamic forces (in these directions) are denoted X_K, X_N, and X_T. These are defined, per *single* ion or molecule, just as in Eqs. (2.2), (2.3), and (2.22). Thus, for K^+ and Na^+,

$$X_K = \mu_{K_o} - \mu_{K_i}, \quad X_N = \mu_{N_i} - \mu_{N_o}. \tag{2.28}$$

Note the reversal of subscripts, out and in. The membrane potential is not included, as already mentioned. X_K and X_N are defined in the positive cycle directions (Fig. 1.6) but they have negative values. That is, as pointed out in Section 1, cycles a and b would run spontaneously in the negative (clockwise) direction (J_a, $J_b < 0$).

In this hypothetical model, Eq. (2.27) would apply to each of the six cycles in the diagram but the cycle forces, in relation to operational forces, contain stoichiometric integers:

$$X_a = 2X_K, \quad X_b = 3X_N, \quad X_c = X_T, \quad X_d = 2X_K + 3X_N$$

$$X_f = 3X_N + X_T, \quad X_g = 2X_K + 3X_N + X_T. \tag{2.29}$$

Of these, in normal operation, X_c, X_f, and X_g are positive (X_T is the dominating force), whereas X_a, X_b, and X_d are negative.

Free energy transduction in this system was discussed at the end of Section 1. It will be obvious from all the examples of the present section that at least two conditions are necessary in order to have free energy transduction: (a) the system must involve two or more operational forces; and (b) there must be at least one cycle in the diagram that includes two or more operational forces. Only in such a cycle can transduction occur.

3. Operational, Cycle, and Transition Fluxes

We have already introduced, in Section 1, the one-way cycle fluxes $J_{\kappa+}$ and $J_{\kappa-}$, and the net cycle flux (+ direction) $J_\kappa = J_{\kappa+} - J_{\kappa-}$, for any cycle κ in a system at steady state. For example, $J_{\kappa+}$ is the mean number of κ cycles completed in the positive direction per unit time, for a single system of the ensemble, during a very long period of time (in which the system does a random walk on its kinetic diagram).

Another kind of flux, called a transition flux, can be defined along any line ij of the kinetic diagram. Thus, a transition flux is more "local" than a cycle flux. The net transition flux in the direction $i \rightarrow j$, along the line ij, is defined as

$$J_{ij} = \alpha_{ij}p_i - \alpha_{ji}p_j, \quad J_{ji} = -J_{ij}, \tag{3.1}$$

where p_i and p_j are the steady-state probabilities already introduced. Obviously [Eq. (2.6)], $J_{ij} = 0$ at equilibrium. J_{ij} is the difference between two one-way transition fluxes. Some net transition fluxes are directly and easily observable, because of a particular change in state of E. For example, in Fig. 1.6, if inorganic phosphate is released in the transition $8 \rightarrow 6$, measurement of the net rate of appearance of phosphate would give J_{86}.

Each operational thermodynamic force (Section 2) has a conjugate (i.e., corresponding) operational flux, with positive direction chosen in the same way for both. For example, in Fig. 1.2, J_M and J_L are the net rate of appearance of M and L (number per unit time) on the outside, per complex E, or the net rate of disappearance on the inside, per complex (the four concentrations are maintained constant). Both J_M and J_L are positive in the system as described in Section 1.

As already implied, operational fluxes are usually related in a simple way to transition fluxes (in complicated diagrams, more than one transition flux might contribute additively to the same operational flux). In the noncomplicated diagram of Fig. 1.2, at steady state,

$$J_M = J_{13} = J_{42} = J_{21}$$

$$J_L = J_{35} = J_{64} = J_{56}. \tag{3.2}$$

The equalities of transition fluxes here are physically obvious but can be deduced formally from the kinetic differential equations (see Sections 6 and 7). Incidentally, J_{21} and J_{56} would be difficult to measure directly and independently.

All of the steady-state flux activity of a system can also be accounted for in terms of cycle fluxes: each completed cycle generally produces some change or changes in the small molecules (some cycles, in some diagrams, accomplish nothing). Hence, operational fluxes can be written as the sum of those cycle fluxes that contribute to the various processes. Individual cycle fluxes are, in general, not observable. Similarly, a transition flux J_{ij} can be written as the sum of cycle fluxes for those cycles that include the line ij (with due regard for algebraic sign). In Fig. 1.2,

$$J_M = J_a + J_c = J_{13} = J_{42} = J_{21}$$

$$J_L = J_b + J_c = J_{35} = J_{64} = J_{56}. \tag{3.3}$$

Also,

$$J_{34} = J_a - J_b. \tag{3.4}$$

In the system as described in Section 1, J_a and J_c are positive and J_b is negative.

To summarize, there are three interdependent levels of fluxes in systems represented by diagrams with cycles: the individual transition fluxes; the cycle fluxes; and operational fluxes (the least detailed set). In principle, at least, transition fluxes and operational fluxes are observable, but not individual cycle fluxes. There are one-way subdivisions (see Sections 7 and 8) of all three kinds of flux. In complicated diagrams, there can be more (nonzero) cycle fluxes than transition fluxes (see Section 7). Hence the former cannot be deduced from knowledge of the latter though the opposite is possible. (However, cycle fluxes cannot be measured directly.)

Flux Relations in Fig. 1.6

The operational and cycle thermodynamic forces for this model were considered in Section 2. Here we examine the three kinds of fluxes involved. The positive direction for K^+ (two ions per cycle) is out $\rightarrow$ in; the positive direction for Na^+ (three ions per cycle) is in $\rightarrow$ out. Examination of Fig. 1.6(b) makes it clear, then, that

$$J_K = 2(J_a + J_d + J_g)$$

$$J_N = 3(J_b + J_d + J_f + J_g) \qquad (3.5)$$

$$J_T = J_c + J_f + J_g.$$

Cycles a, d, and g appear in the first equation because these are the only cycles that contribute to K^+ transport, etc. From section 1, J_a, J_b, J_d are negative whereas J_c, J_f, and J_g are positive. Note that, compared to Eqs. (2.29), the stoichiometric numbers are reversed, so to speak: here they multiply *cycle* fluxes; in Eqs. (2.29) they multiply *operational* forces. Also, independent variables in these linear equations are reversed: here the independent variables are *cycle* fluxes; in Eqs. (2.29) they are *operational* forces. These relationships will prove to be important in Section 5.

From Fig. 1.6(a), we see that the relations between cycle and transition fluxes are:

$$J_{42} = J_{21} = J_{31} = J_a + J_d + J_g = J_K/2$$

$$J_{35} = J_{64} = J_b + J_d + J_f + J_g = J_N/3$$

$$J_{57} = J_{78} = J_{86} = J_c + J_f + J_g = J_T \qquad (3.6)$$

$$J_{34} = J_a - J_b - J_f$$

$$J_{56} = J_b + J_d - J_c.$$

All three kinds of flux, including their one-way decompositions, for any kinetic diagram, can (at least in principle) be expressed as algebraic functions of all the first-order rate constants of the diagram. These relations will be discussed in Sections 6 and 7.

Relation Between $J_{\kappa\pm}$ and $\Pi_{\kappa\pm}$

For completeness, we anticipate here a simple and intuitively plausible relation between $J_{\kappa\pm}$ and $\Pi_{\kappa\pm}$. This relation will be established in Section 7. We want to mention it here, in advance, because of its resemblance to Eq. (2.27) and also because it will be needed in Section 5.

During the course of a long random walk (Section 1) on the lines of a kinetic diagram, any particular cycle κ of the diagram will be completed at a mean rate $J_{\kappa+}$ in the positive direction and at a mean rate $J_{\kappa-}$ in the negative direction. The walk takes place over the whole diagram, so naturally $J_{\kappa+}$ and $J_{\kappa-}$ will depend on *all* of the α_{ij} (first-order rate constants) of the diagram. This dependence will be made explicit in Section 7. However, it is reasonable to guess that $J_{\kappa+}$ and $J_{\kappa-}$ might be proportional to $\Pi_{\kappa+}$ and $\Pi_{\kappa-}$, respectively, that is, to the products of rate constants around the κ cycle itself. This, in fact, turns out to be the case (Section 7). Furthermore, not surprisingly, the proportionality constant (which involves all of the rate constants of the diagram) is the same for $J_{\kappa+}$ and $J_{\kappa-}$. Thus, for each cycle κ in any diagram for a system at steady state, there is the simple relation [see Eq. (2.27)]

$$\frac{J_{\kappa+}}{J_{\kappa-}} = \frac{\Pi_{\kappa+}}{\Pi_{\kappa-}} = e^{X_\kappa/kT}. \tag{3.7}$$

All of the quantities in this equation are always positive except X_κ, which may be positive or negative. At equilibrium,

$$J_{\kappa+} = J_{\kappa-}, \quad \Pi_{\kappa+} = \Pi_{\kappa-}, \quad X_\kappa = 0, \quad J_\kappa = 0. \tag{3.8}$$

Cycle completions occur in the random walk at equilibrium as well as at steady state.

Because $J_\kappa = J_{\kappa+} - J_{\kappa-}$, an important consequence of Eq. (3.7) is that J_κ and X_κ always have the same sign. That is, for any cycle, a statistical excess of cycle completions will occur in the direction of the thermodynamic force driving the cycle. This is what one should expect, in view of the second law of the thermodynamics.

4. Efficiency and the Rate of Free Energy Dissipation

An ensemble of complexes, operating at constant temperature and pressure and at a nonequilibrium steady state, carries out its net processes (transport, chemical reactions) spontaneously. According to the second law of thermodynamics, there must be accompanying dissipation of Gibbs free energy. We designate the rate of free energy dissipation per complex or system by Φ. For example, in Fig. 1.4(b), for each net cycle completion ($+$ direction) by one enzyme molecule E, one S is converted to one P in solution. The associated free energy decrease is $X = \mu_S - \mu_P$ [Eq. (2.13)]. Let J be the net rate of cycle completions ($+$ direction) per enzyme molecule. Then, in this example, $\Phi = JX$. J and X have the same sign [Eq. (3.7)] so $\Phi > 0$, as required by the second law. At equilibrium, $X = 0$, $J = 0$, and $\Phi = 0$.

Let us turn now to the more complicated case, Fig. 1.2, where there are two operational forces and fluxes, and three cycles. As already seen, the total steady-state net activity of this kind of system can be represented at any one of three different levels: operational fluxes, cycle fluxes, or transition fluxes. We consider the first two representations here but defer the third until Chapter 3 where necessary transition free energy changes will be introduced.

At the operational level, in Fig. 1.2, for each M or L molecule transported from inside to outside, the free energy drop is X_M [Eq. (2.2)] or X_L [Eq. (2.3)], respectively. The net rates, per complex, at which these events occur are J_M and J_L, respectively. Hence, the rate of free energy dissipation is

$$\Phi = J_M X_M + J_L X_L \geq 0. \tag{4.1}$$

The second law requires that $\Phi > 0$ at a nonequilibrium steady state.

24

It should be recalled that J_M, J_L, and X_M are positive (in normal operation) but X_L is negative. Hence $J_M X_M$ must be larger than $J_L(-X_L)$. We shall return to this point below, in connection with the efficiency.

The total net activity in the same system can also be represented at the cycle level. For each net completion of cycle κ (κ = a, b, c) in the + direction, the free energy decrease is X_κ. The net rate of these cycle completions is J_κ. Hence Φ can also be written as

$$\Phi = J_a X_a + J_b X_b + J_c X_c \geq 0. \tag{4.2}$$

Because J_κ and X_κ always have the same sign [Eq. (3.7)], each separate cycle term $J_\kappa X_\kappa$ makes a positive contribution to Φ (except at equilibrium where $X_\kappa = 0$, $J_\kappa = 0$, and $\Phi = 0$). The cycle level is more detailed than the operational level. The transition level is still more detailed; again it is found (see Chapter 3), as in Eq. (4.2), that each separate transition term in Φ is positive.

Equations (4.1) and (4.2) have to be consistent with each other. This is easily confirmed using Eqs. (2.5) and (3.3):

$$\begin{aligned}
\Phi &= J_M X_M + J_L X_L \\
&= (J_a + J_c) X_M + (J_b + J_c) X_L \\
&= J_a X_M + J_b X_L + J_c(X_M + X_L) \\
&= J_a X_a + J_b X_b + J_c X_c.
\end{aligned} \tag{4.3}$$

As another example, consider Fig. 1.6 and Eqs. (2.29) and (3.5). The expression for Φ at the operational level is

$$\Phi = J_K X_K + J_N X_N + J_T X_T \geq 0. \tag{4.4}$$

Under normal conditions, J_K, J_N, J_T, and X_T are positive but X_K and X_N are negative. Hence $J_T X_T$ is larger than $J_K(-X_K) + J_N(-X_N)$. At the cycle level,

25

$$\Phi = J_a X_a + J_b X_b + J_c X_c + J_d X_d + J_f X_f + J_g X_g \geq 0, \quad (4.5)$$

where all terms are positive (or zero). The equivalence of Eqs. (4.4) and (4.5) is seen from

$$\Phi = J_K X_K + J_N X_N + J_T X_T$$

$$= 2(J_a + J_d + J_g)X_K + 3(J_b + J_d + J_f + J_g)X_N$$

$$+ (J_c + J_f + J_g)X_T \qquad (4.6)$$

$$= J_a(2X_K) + J_b(3X_N) + J_c X_T + J_d(2X_K + 3X_N)$$

$$+ J_f(3X_N + X_T) + J_g(2X_K + 3X_N + X_T).$$

In view of Eqs. (2.29), this is the same as Eq. (4.5).

Efficiency of Free Energy Transduction

Again we turn to Figs. 1.2 and 1.6 to illustrate some general principles. In the Fig. 1.2 system, M drives L against its concentration gradient. That is, part of the free energy drop in M, as the ensemble of complexes operates at steady state, is used to *increase* the free energy of L: not all of the free energy drop in M is dissipated; some of it is transferred to L. This is an example of what is meant by free energy transfer or transduction. The rate of free energy loss by M, per complex, is $J_M X_M$. The part of this loss that is transferred to L (not wasted) is $J_L(-X_L)$. The remainder,

$$J_M X_M - J_L(-X_L),$$

is dissipated, as already stated in Eq. (4.1). The efficiency of the conversion of M free energy into L free energy is obviously

$$\eta = \frac{J_L(-X_L)}{J_M X_M}. \qquad (4.7)$$

26

The efficiency becomes unity only at equilibrium, where the two fluxes, the two forces, and Φ [Eq. (4.1)] are all zero. Otherwise, $\eta < 1$ because $\Phi > 0$.

Using Eqs. (3.3), Eq. (4.7) can be rewritten as

$$\eta = \frac{(J_b + J_c)(-X_L)}{(J_a + J_c)X_M}. \tag{4.8}$$

Here, it will be recalled, J_b is negative whereas J_a and J_c are positive. The free energy transduction occurs in cycle c. It will be noticed that a negative J_b and a positive J_a both reduce the efficiency. Cycles a and b are used only when there is slippage (states $3 \rightleftarrows 4$). Thus, as we might have anticipated, slippage reduces the efficiency. When there is no slippage, $J_a = J_b = 0$ and

$$\eta = \frac{-X_L}{X_M} \quad \text{(cycle c only).} \tag{4.9}$$

In each completion of cycle c in the $+$ direction, the free energy of M decreases by an amount X_M while the free energy of L increases by an amount $-X_L$. The amount of free energy dissipation per cycle is $X_M - (-X_L)$, or X_c, as already mentioned in Section 2.

In the model of Fig. 1.6, at steady state, Φ in Eq. (4.4) can be written as

$$\Phi = J_T X_T - J_K(-X_K) - J_N(-X_N) > 0, \tag{4.10}$$

where the last two terms are negative. Part of the free energy decrease of ATP is used to increase the free energy of K^+ and Na^+ by driving both across the membrane against their concentration gradients. The efficiency of this free energy transfer is

$$\eta = \frac{J_K(-X_K) + J_N(-X_N)}{J_T X_T}. \tag{4.11}$$

If we introduce cycle fluxes,

27

$$\eta = \frac{2(J_a + J_d + J_g)(-X_K) + 3(J_b + J_d + J_f + J_g)(-X_N)}{(J_c + J_f + J_g)X_T}. \quad (4.12)$$

The fluxes J_a, J_b, and J_d are negative; J_c and J_f are positive. All of these arise from slippage ($3 \rightleftarrows 4$ and $5 \rightleftarrows 6$). It is obvious from the form of Eq. (4.12) that cycles a, b, d (negative in the numerator), and c (positive in the denominator) all reduce the efficiency. The effect of cycle f (ATP is used to transport Na^+) is not so clear because J_f occurs in both numerator and denominator. However, if J_f is small and J_g large, it is not hard to show that J_f in Eq. (4.12) also reduces η.

When there is no slippage at all, only cycle g is used: there is complete coupling of K^+ and Na^+ transport with the ATPase activity. In this case Eq. (4.12) becomes

$$\eta = \frac{2(-X_K) + 3(-X_N)}{X_T} \quad \text{(cycle g only).} \quad (4.13)$$

The stoichiometry in this case is exactly $2:3:1$. This is spoiled when other cycles participate.

Free Energy Transduction

The examples we have used illustrate how free energy can be transferred from one small molecule to another (or to more than one). This can be accomplished, however, only with the participation of a macromolecular protein complex or enzyme. The complex itself does not change in the overall process; it serves as a cycling interchange depot for the small molecules.

The rate constants α_{ij} of the kinetic diagram are related to the thermodynamics of the small molecule changes (Section 2). Obviously, these rate constants are related even more strongly to the kinetics of the processes involved (Section 3), though the details of the connections between the α_{ij} and the various fluxes will not be considered until Chapter 2. The rate constants of the kinetic diagram not only have to conform to the thermodynamic requirements of the free energy transduction, they also must have values such that the necessary cyclic processes occur at a sufficiently rapid rate.

It should be emphasized again that free energy transduction requires that the kinetic diagram must have at least one cycle that involves *both* the free energy donor and the free energy acceptor; only in such cycles can transduction occur. Examples are cycle c in Fig. 1.2 and cycle g in Fig. 1.6. Free energy transduction is clearly a property of complete cycles and not of particular individual transitions within a cycle; *every* transition in a transduction cycle is crucial as a participant in the mechanism, in the cycle force, and in the cycle flux.

5. Fluxes and Forces Near Equilibrium

When the thermodynamic forces are zero, the steady state reached is an equilibrium state: all the net mean fluxes are also zero. (However, as already mentioned, the one-way cycle fluxes are not zero.) If the forces are arranged to be small but nonzero, in response the net mean fluxes will also have small nonzero values. (In special cases, some but not all forces and fluxes might be zero.) In fact, in this near-equilibrium regime (small forces), the fluxes will depend *linearly* on the forces. For example, if f is a flux that depends on two forces x and y, expansion of $f(x, y)$ about $x = y = 0$ gives

$$f(x, y) = \left(\frac{\partial f}{\partial x}\right)_{\substack{x=0 \\ y=0}} x + \left(\frac{\partial f}{\partial y}\right)_{\substack{x=0 \\ y=0}} y + \cdots . \tag{5.1}$$

The constant term is missing here because $f = 0$ when $x = y = 0$. Negligible quadratic and higher terms in the expansion are omitted. The derivatives are properties of the equilibrium state ($x = y = 0$), so the dependence of f on x and y is indeed linear if x and y are small enough.

Thus there is a linear flux-force regime near equilibrium. The theoretical treatment of such systems is often classified as "irreversible thermodynamics" or "nonequilibrium thermodynamics." This subject is important for two reasons. First, general nonequilibrium theory[5] is notoriously difficult; first (linear) departures from equilibrium are obviously the easiest place to start in approaching nonequilibrium problems. Lars Onsager[6] put irreversible thermodynamics (the linear regime) on a firm foundation in 1931. Second, the linear flux-force regime is found, experimentally, to persist in many cases over a surprisingly large range in forces. Hence the subject holds a lot more than merely academic or theoretical interest. The extent of linear regimes in real biophysical systems is a topic covered in detail by Caplan and Essig.[2]

We are interested in this section in any system, near equilibrium, that can be represented by a biochemical kinetic diagram (Onsager's treatment is much broader than this). However, the model in Fig. 1.6 is sufficiently complicated to bring out all the main points, so we base the discussion on this example. Further details and a more general treatment will be found in refs. 7 and 8.

We begin with Eqs. (3.5):

$$J_K = 2(J_a + J_d + J_g)$$

$$J_N = 3(J_b + J_d + J_f + J_g) \tag{5.2}$$

$$J_T = J_c + J_f + J_g.$$

Our object is to put the right-hand sides of these equations, near equilibrium, into the form of linear terms in X_K, X_N, and X_T. The resulting equations will correspond to Eq. (5.1) for this model.

From Eq. (3.7) (which will be verified in Section 7), for any cycle $\kappa = a, b, \ldots,$

$$\frac{J_{\kappa+}}{J_{\kappa-}} = e^{X_\kappa/kT} \tag{5.3}$$

$$J_\kappa = J_{\kappa+} - J_{\kappa-} = J_{\kappa-}(e^{X_\kappa/kT} - 1). \tag{5.4}$$

At equilibrium,

$$J_\kappa^e = 0, \quad J_{\kappa+}^e = J_{\kappa-}^e \equiv J_{\kappa\pm}^e. \tag{5.5}$$

That is, the two one-way cycle fluxes are equal (but nonzero) at equilibrium, and we denote both by $J^e_{\kappa\pm}$. Near equilibrium, where X_κ/kT is small, we expand the right-hand side of Eq. (5.4) in powers of X_κ/kT and keep only the linear term:

$$J_\kappa = J^e_{\kappa\pm}(X_\kappa/kT) = I_\kappa X_\kappa, \qquad (5.6)$$

where we have defined I_κ (to simplify notation) as

$$I_\kappa \equiv J^e_{\kappa\pm}/kT. \qquad (5.7)$$

I_κ is the equilibrium one-way cycle flux for cycle κ, divided by kT. $J^e_{\kappa\pm}$ in Eq. (5.6) is the leading (constant) term in an expansion of $J_{\kappa-}$ in powers of X_κ/kT (see Section 7). $J^e_{\kappa\pm}$, for any cycle κ and any diagram, is an explicit function of an equilibrium set of rate constants for the diagram. Such functions will be illustrated in Section 7. Thus I_κ is a well-defined equilibrium kinetic property for each cycle κ.

If we now substitute Eqs. (2.29) into Eq. (5.6) for each cycle in the diagram, we obtain

$$J_a = 2I_a X_K, \quad J_b = 3I_b X_N, \quad J_c = I_c X_T$$

$$J_d = I_d(2X_K + 3X_N), \quad J_f = I_f(3X_N + X_T) \qquad (5.8)$$

$$J_g = I_g(2X_K + 3X_N + X_T).$$

The final step is to substitute Eqs. (5.8) into Eqs. (5.2). The result can be written in the form

$$J_K = L_{KK}X_K + L_{KN}X_N + L_{KT}X_T$$

$$J_N = L_{NK}X_K + L_{NN}X_N + L_{NT}X_T \qquad (5.9)$$

$$J_T = L_{TK}X_K + L_{TN}X_N + L_{TT}X_T,$$

where

$$L_{KK} = 4(I_a + I_d + I_g)$$

$$L_{NN} = 9(I_b + I_d + I_f + I_g) \tag{5.10}$$

$$L_{TT} = 1(I_c + I_f + I_g)$$

and

$$L_{KN} = L_{NK} = 6(I_d + I_g)$$

$$L_{KT} = L_{TK} = 2I_g \tag{5.11}$$

$$L_{NT} = L_{TN} = 3(I_f + I_g).$$

The notation in Eqs. (5.9) is conventional: the operational fluxes (J_i) are linear functions of the operational forces (X_j) and the coefficients are denoted L_{ij}. The L_{ij} are properties of the equilibrium state (they do not depend on the forces), and comprise a square matrix. The magnitudes of the coefficients L_{ij} determine the magnitudes of the fluxes J_i in response to the imposition of the forces X_j.

The explicit expressions for the L_{ij} in Eqs. (5.10) and (5.11) follow a pattern that obtains for almost all kinetic diagrams:

(1) All of the L_{ij} are simple linear combinations of the I_κ. That is, aside from the factor kT, the L_{ij} are determined by the equilibrium one-way cycle fluxes of the diagram.

(2) The matrix L_{ij} is symmetrical: $L_{ij} = L_{ji}$ (this is sometimes referred to as a "reciprocal relation"). Onsager[6] established this important symmetry in a more general theoretical context; it has been verified experimentally in a number of cases. Physically, the effect of the force X_j on the flux J_i is the same as the effect of the force X_i on the flux J_j.

(3) Comparison of Eqs. (5.10) and Fig. 1.6 shows that a diagonal L_{ii} for operational process i has an equilibrium one-way cycle flux contribution from each cycle in the diagram that includes process i, but from no other cycles. The numerical coefficients in Eqs. (5.10) are squares of the stoichiometric numbers $2:3:1$.

(4) Comparison of Eqs. (5.11) and Fig. 1.6 shows that a nondiagonal L_{ij} has an equilibrium one-way cycle flux contribution from those cycles and only those cycles that include *both* process i and process j. Here is the origin of the symmetry of the L_{ij} matrix. The numerical coefficients in Eqs. (5.11) are products of the associated stoichiometric numbers: $2 \cdot 3$, $2 \cdot 1$, and $3 \cdot 1$ for KN, KT, and NT, respectively.

With the above pattern in mind, when confronted with another diagram, one can express the L_{ij} in terms of the I_κ simply by inspection of the diagram and its cycles, without repeating the algebra in Eqs. (5.2) and (5.8). For example, from Fig. 1.2 (the stoichiometric numbers are unity),

$$J_M = (I_a + I_c)X_M + I_c X_L$$

$$J_L = I_c X_M + (I_b + I_c)X_L. \tag{5.12}$$

As already mentioned, to go further and write the I_κ as functions of the diagram rate constants requires the use of the methods in Section 7.

Reference was made above, prior to the listing of items (1) to (4), to "almost all kinetic diagrams." Exceptions, where the simple sort of algebra in Eqs. (5.2) and (5.8) would have to be repeated on a case by case basis, are diagrams in which the stoichiometric number for some operational process i (or more than one such process) is not the same in every cycle that includes process i.

Coupling

The presence of nondiagonal terms in Eqs. (5.9) and (5.12) indicates that the operational flux J_i is influenced not only by the force X_i, but also by the other operational forces. In other words, the operational processes are "coupled" to each other. Such coupling is essential for free energy transduction: one process cannot influence or drive another process if the two processes are independent. Item (4), above, emphasizes again that: (a) coupling and free energy transduction require that the kinetic diagram must contain at least one cycle that involves two or more operational forces and fluxes; and (b) cyclic activity is the essential feature in free energy transduction.

The L_{ij} in Eqs. (5.9) are independent of the values selected for the X_j. For example, let us choose $X_K \neq 0$, $X_N = 0$, $X_T = 0$. Then $J_K = L_{KK} X_K$. Also, in this case, from Eq. (4.4), $J_K X_K > 0$. That is, J_K and X_K have the same sign. Hence L_{KK} must be positive. This same argument can be applied to any operational process in any diagram. Hence, all diagonal L_{ii} are positive: when only the force X_i is imposed on the system, the resulting mean flux J_i will always be in the direction of the force (a consequence of the second law of thermodynamics).

The nondiagonal L_{ij} may be positive or negative. They are all positive in Eqs. (5.9) and (5.12) because all cycle and operational fluxes were chosen as positive in the same direction (counterclockwise) in Figs. 1.2 and 1.6. This choice need not be made.

If, hypothetically, all the nondiagonal $L_{ij} = 0$, then the different operational processes would be completely uncoupled (i.e., independent of each other). For example, in Eqs. (5.9),

$$J_K = L_{KK} X_K, \quad J_N = L_{NN} X_N, \quad J_T = L_{TT} X_T. \tag{5.13}$$

Complete coupling is the other extreme. For example, in Eqs. (5.9), suppose only cycle g is significant (no slippage). Then

$$J_K = 4I_g X_K + 6I_g X_N + 2I_g X_T = 2I_g X_g = 2J_T$$

$$J_N = 6I_g X_K + 9I_g X_N + 3I_g X_T = 3I_g X_g = 3J_T \tag{5.14}$$

$$J_T = 2I_g X_K + 3I_g X_N + I_g X_T = I_g X_g.$$

The fluxes are coupled by use of the single cycle and are in the exact stoichiometric ratio, $2:3:1$. This same ratio is also found, far from equilibrium, when only cycle g is active [Eqs. (3.5)]: every completed cycle g moves $2K^+$, $3Na^+$, and hydrolyzes 1 ATP. As another example, of complete coupling, if only cycle c contributes in Eqs. (5.12), we have

34

$$J_M = I_c X_M + I_c X_L = I_c X_c$$

$$J_L = I_c X_M + I_c X_L = I_c X_c = J_M. \tag{5.15}$$

The two fluxes are necessarily equal in this special case because of the use of cycle c only.

More generally, for any two operational processes i and j, we can define the degree of coupling,[2] near equilibrium, by

$$q = L_{ij}/(L_{ii}L_{jj})^{1/2}. \tag{5.16}$$

In Eq. (5.13), where there is no coupling at all, $q = 0$ for each of the three pairs because the $L_{ij} = 0$. In Eqs. (5.14) and (5.15), $q = 1$ for every pair. It is easy to see[2] that the limits for q are ± 1 (either of these limits corresponds to complete or tight coupling). In other words,

$$q^2 \leq 1, \quad L_{ij}^2 \leq L_{ii}L_{jj}. \tag{5.17}$$

When the coupling is less than complete, the inequalities hold. The latter inequality, for any kinetic diagram, can be seen to be a direct consequence of items (3) and (4), above: the cycles that contribute to L_{ij} necessarily also contribute to both L_{ii} and L_{jj}, but usually there will be additional cycles that include process i or process j but *not* both. These cycles contribute to L_{ii} or L_{jj} but *not* to L_{ij}. These extra contributions to L_{ii} or L_{jj} produce the inequalities in Eqs. (5.17) and less than complete coupling. Examples, from Eqs. (5.9) and (5.12) are:

$$\begin{cases} L_{KN}^2 = 36(I_d + I_g)^2 \\ \\ L_{KK}L_{NN} = 36(I_d + I_g + \underline{I_a})(I_d + I_g + \underline{I_b} + \underline{I_f}) \end{cases} \tag{5.18}$$

$$L_{KT}^2 = 4I_g^2, \quad L_{KK}L_{TT} = 4(I_g + \underline{I_a} + I_d)(I_g + \underline{I_c} + I_f) \tag{5.19}$$

$$\begin{cases} L_{NT}^2 = 9(I_f + I_g)^2 \\ \\ L_{NN}L_{TT} = 9(I_f + I_g + \underline{I_b} + \underline{I_d})(I_f + I_g + \underline{I_c}) \end{cases} \tag{5.20}$$

$$L_{ML}^2 = I_c^2, \quad L_{MM}L_{LL} = (I_c + \underline{I_a})(I_c + \underline{I_b}). \tag{5.21}$$

In every case the underlined extra terms in $L_{ii}L_{jj}$ make this product larger than L_{ij}^2 (unless all extra terms have the value zero; that is, unless there is no slippage).

Of course the concepts of complete (tight) coupling, via use of a single cycle, or of noncoupling, still have significance at steady states arbitrarily far from equilibrium, but the quantitative parameter q is defined only for the linear regime near equilibrium.

The rate of free energy dissipation in Eq. (4.4) can be put in more explicit form in the linear regime near equilibrium. Substitution of Eqs. (5.9) for the J_i yields

$$\Phi = L_{KK}X_K^2 + L_{NN}X_N^2 + L_{TT}X_T^2 + 2L_{KN}X_KX_N + 2L_{KT}X_KX_T$$

$$+ 2L_{NT}X_NX_T \geq 0. \tag{5.22}$$

This is a quadratic form in the forces. If X_K and X_N are negative (as they are, normally, far from equilibrium), the first four terms are positive and the last two terms are negative. The sum must be positive (or zero, at equilibrium). Alternatively, if we put $J_\kappa = I_\kappa X_\kappa$ for each cycle κ in Eq. (4.5), we have

$$\Phi = I_aX_a^2 + I_bX_b^2 + I_cX_c^2 + I_dX_d^2 + I_fX_f^2 + I_gX_g^2 \geq 0. \tag{5.23}$$

Here, every term is positive or zero: the second law applies to the mean steady-state activity in each cycle separately (see Section 4). It is easy to verify that Eqs. (5.22) and (5.23) are equivalent.

Summary of Chapter 1

In order for one small molecule (the donor) to transfer free energy steadily to one or more other small molecules (the acceptor or acceptors), it is apparently necessary to use a large enzyme, protein, or protein complex as intermediary. The complex, in its interactions with the small molecules, can exist in a number of discrete states, with possible reversible transitions between some pairs of these states. The states and transitions can be represented by a diagram or graph, including a first-order rate constant α_{ij} for each possible transition $i \to j$ between states. The diagram must include at least one cycle in order for there to be nonequilibrium steady-state activity in the system. In order for the complex to accomplish its job of facilitating free energy transfer from one small molecule to another, the diagram must contain at least one cycle that includes the net processes of both the free energy donor and the free energy acceptor. Such a cycle represents a repetitive finite sequence of states passed through by the complex, in its interactions with donor and acceptor, that accomplishes the required changes in donor and acceptor but leaves the complex itself unchanged. The thermodynamic force driving the steady cycling is the excess free energy change in the donor in the cyclic process (i.e., excess over the amount of free energy transferred to the acceptor). Even with favorable thermodynamics driving the cycle, the steady state cycling must, to be useful, occur at a sufficiently high rate (flux). This rate will obviously depend on the rate constants α_{ij} of the diagram (the α_{ij} must also be consistent with the required thermodynamics). The relations between fluxes and rate constants are discussed in Sections 6 and 7.

References

1. Hill, T. L. (1977) *Free Energy Transduction in Biology* (Academic, New York).
2. Caplan, S. R. and Essig, A. (1983) *Bioenergetics and Linear Nonequilibrium Thermodynamics* (Harvard, Cambridge).
3. Westerhoff, H. V. and van Dam, K. (1987) *Thermodynamics and Control of Biological Free Energy Transduction* (Elsevier, Amsterdam).
4. Hill, T. L. and Eisenberg, E. (1981) Q. Revs. Biophys. **14**, 463.
5. Keizer, J. (1987) *Statistical Thermodynamics of Nonequilibrium Processes* (Springer-Verlag, New York).
6. Onsager, L. (1931) Phys. Rev. **37**, 405.
7. Hill, T. L. (1982) Nature **299**, 84.
8. Hill, T. L. (1983) Proc. Natl. Acad. Sci. USA **80**, 2589.

2

State Probabilities and Fluxes in Terms of the Rate Constants of the Diagram

The basic principles of free energy transduction were introduced in Chapter 1. Steady-state probabilities and especially forces and fluxes played a prominent role in the discussion. One important topic was bypassed, however, which is covered in the present chapter: How are the state probabilities and the fluxes related to the first-order rate constants of α_{ij} of the kinetic diagram? This subject does not add much to the general concept of free energy transduction *per se*, but it is essential to an understanding of the steady-state kinetic activity of the cyclic system that carries out the free energy transduction. Section 6 deals with state probabilities (and fluxes in very simple systems) and Section 7 is concerned with fluxes and related kinetic topics. Section 8 is essentially an appendix on recent and more advanced topics related to cycles and diagrams.

6. The Diagram Method for State Probabilities

We introduce this subject by means of an example of optimal complexity for pedagogical purposes. To be explicit, we use the model

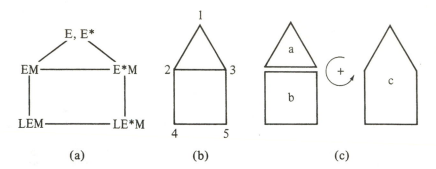

Fig. 2.1. (a) Model used to illustrate the diagram method. (b) Diagram. (c) Cycles.

in Fig. 1.2(a) but assume that there is a fast equilibrium between states E and E* (to simplify the algebra somewhat). This leads to the diagram and cycles in Fig. 2.1, with renumbered states. State 1 is the equilibrium mixture of E and E*.

The cyclic activity in Fig. 2.1 is the topic of primary interest, but we postpone this subject until Section 7. Here we ask a simpler question: Given all the first-order rate constants α_{ij} for the diagram in Fig. 2.1(b) (there are 6 lines in the diagram and, hence, 12 different α_{ij}), what is the probability p_i that each state $i = 1, \ldots, 5$ is occupied at steady state? That is, what fraction of time does a given complex spend in each state? An equivalent question is: Given a large ensemble of these complexes at steady state, what fraction of them are in state i at an arbitrary time? The straightforward way to answer this question is to start with the differential equations for the p_i. These are obvious from an inspection of the diagram:

$$\frac{dp_1}{dt} = J_{21} + J_{31} = (\alpha_{21}p_2 - \alpha_{12}p_1) + (\alpha_{31}p_3 - \alpha_{13}p_1)$$

$$(6.1)$$

$$\frac{dp_2}{dt} = J_{12} + J_{32} + J_{42} = (\alpha_{12}p_1 - \alpha_{21}p_2) + (\alpha_{32}p_3 - \alpha_{23}p_2)$$

$$+ (\alpha_{42}p_4 - \alpha_{24}p_2),$$

etc. There are five of these equations but only four are independent because of the normalization requirement

$$p_1 + p_2 + p_3 + p_4 + p_5 = 1. \qquad (6.2)$$

At steady state, the $dp_i/dt = 0$. In this case we can use Eq. (6.2) and any four of Eqs. (6.1) to give five linear nonhomogeneous algebraic equations in five unknowns, the p_i. If only numerical results are needed in a special case (i.e., numerical values are assigned to the α_{ij}), the simplest procedure is to solve the equations by computer (matrix inversion). But if one wants to see how each p_i depends explicitly on all the α_{ij}, the standard (tedious) method would be to solve the linear equations using Cramer's rule. There is, however, an alternative graphical or diagram method that is more elegant mathematically and can be understood intuitively. This method has apparently been "discovered" a number of times (starting in the last century), most recently by King and Altman[1] and by Hill.[2,3] A relatively simple proof is available [2-4] but will not be included here. Actually, the main reason for including this diagram method for state probabilities is that it can be extended[2,3] to deduce cycle fluxes (Section 7), which are more important and interesting in the present context.

In applying the diagram method to Fig. 2.1(b), the first step is to construct the complete set of *partial diagrams*, each of which contains the maximum possible number of lines (four here) that can be included in the diagram without forming any cycle (closed path). There are 11 such partial diagrams in this case, shown in Fig. 2.2. If one more line is introduced into any vacant position in any of these partial diagrams, a cycle is produced. At least one line goes to each vertex (state) in a partial diagram (otherwise more lines could be introduced without forming a cycle).

The next step is to introduce arrows (i.e., a directionality for each line) into the partial diagrams of Fig. 2.2 in five different ways, one way for each state (vertex). For example, consider state 2. Figure 2.3 shows the eleven *directional diagrams* for this state, as obtained from Fig. 2.2. The algorithm for introducing arrows is simple: all connected paths in Fig. 2.3 are made to "flow" *toward* and *end* at vertex 2. It

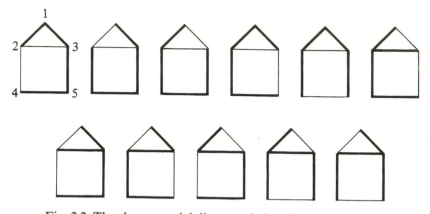

Fig. 2.2. The eleven partial diagrams belonging to Fig. 2.1(b).

will be noted that in the flow toward the ultimate vertex (vertex 2 in Fig. 2.3), "streams" may converge but they never diverge (for this would require a cycle in the partial diagram).

There is a set of eleven directional diagrams for each of the five states. In each case, all streams flow toward—and end at—the particular state being considered.

Now each directional line or arrow in Fig. 2.3 corresponds to a rate constant: an arrow from state i to a neighboring state j corresponds to the rate constant α_{ij}. Thus, each directional diagram in Fig. 2.3 represents a product of four rate constants, one for each arrow, as indicated under the first and third of these diagrams.

Considering each directional diagram as a product of rate constants, the final step is the following: each p_i is proportional to the sum of the directional diagrams belonging to the state i. But, since $\Sigma_i p_i = 1$,

$$p_i = \frac{\text{sum of directional diagrams of state } i}{\text{sum of directional diagrams of } all \text{ states } (\equiv \Sigma)}. \qquad (6.3)$$

Σ includes every rate constant in the diagram. In the above example, there are $5 \times 11 = 55$ directional diagrams altogether. Therefore, for

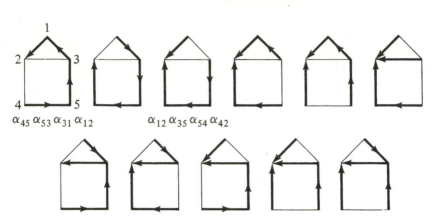

$$\alpha_{45}\,\alpha_{53}\,\alpha_{31}\,\alpha_{12} \qquad\qquad \alpha_{12}\,\alpha_{35}\,\alpha_{54}\,\alpha_{42}$$

Fig. 2.3. Directional diagrams for state 2. Algebraic values of the first and third directional diagrams are given.

state 2, say,

$$p_2 = \frac{(\alpha_{45}\alpha_{53}\alpha_{31}\alpha_{12} + \alpha_{13}\alpha_{35}\alpha_{54}\alpha_{42} + 9 \text{ other terms})}{(\alpha_{45}\alpha_{53}\alpha_{31}\alpha_{12} + \alpha_{13}\alpha_{35}\alpha_{54}\alpha_{42} + 53 \text{ other terms})}. \quad (6.4)$$

The result (6.3) is intuitively reasonable because the steady-state probability of occupation of the ith state is proportional to the sum of products of rate constants along different routes leading *toward* state i. That is, we would expect in a general way that the larger the rate constants leading *toward* state i the larger the relative population of state i at $t = \infty$.

Equation (6.3) is, of course, also the solution for p_i in the special case that the rate constants correspond to equilibrium at $t = \infty$. Incidentally, at equilibrium, each separate pair of terms (in parentheses) in Eqs. (6.1) is equal to zero (i.e., each $J_{ij} = 0$).

Some Simpler Examples

The above example establishes the method. We now apply it to three very simple examples for which it is easy to write explicit results.

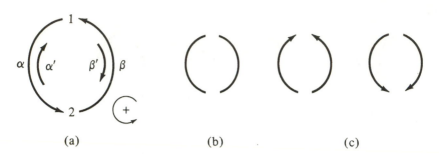

(a)   (b)   (c)

Fig. 2.4. (a) Two-state cycle. (b) The two partial diagrams. (c) The four directional diagrams.

Two-State Cycle. Figure 2.4(a) shows a two-state cycle. This could arise, for example, from Fig. 1.4(b) if EP is a transient intermediate. In this case, $\alpha = \alpha^* c_S$ and $\beta' = \beta^* c_P$ (see Section 2). Often β' is negligible (Michaelis-Menten kinetics). The two partial diagrams for the two-state cycle are shown in Figure 2.4(b) and the four directional diagrams are in Fig. 2.4(c). From Eq. (6.3), then,

$$p_1 = \frac{\alpha' + \beta}{\Sigma}, \quad p_2 = \frac{\alpha + \beta'}{\Sigma}, \quad \frac{p_1}{p_2} = \frac{\alpha' + \beta}{\alpha + \beta'} \tag{6.5}$$

$$\Sigma = \alpha + \alpha' + \beta + \beta'.$$

Note also that (see Section 2)

$$\Pi_+ = \alpha\beta, \quad \Pi_- = \alpha'\beta', \quad e^{X/kT} = \frac{\Pi_+}{\Pi_-} = \frac{\alpha\beta}{\alpha'\beta'}, \tag{6.6}$$

where X is the thermodynamic force in the cycle. This is also the operational force because the diagram has only a single cycle. At equilibrium, there is detailed balance along each line of the cycle:

$$X = 0, \quad \Pi_+ = \Pi_-, \quad \alpha\beta = \alpha'\beta' \tag{6.7}$$

$$\frac{p_1^e}{p_2^e} = \frac{\alpha'}{\alpha} = \frac{\beta}{\beta'} = \frac{\alpha' + \beta}{\alpha + \beta'}.$$

Because the diagram contains only a single cycle, the flux can be discussed here without use of the special method in Section 7. In fact, because of the single cycle, the cycle flux, the operational flux, and the two transition fluxes [Eq. (3.1)] are all equal:

$$J = J_{12(\alpha)} = \alpha p_1 - \alpha' p_2$$

$$= J_{21(\beta)} = \beta p_2 - \beta' p_1$$

$$= \frac{\alpha\beta - \alpha'\beta'}{\Sigma} = \frac{\Pi_+ - \Pi_-}{\Sigma}, \tag{6.8}$$

where the last line follows from Eqs. (6.5). Note that $J = 0$ at equilibrium. Incidentally,

$$\frac{dp_1}{dt} = 0 = (\alpha' p_2 - \alpha p_1) + (\beta p_2 - \beta' p_1)$$

$$= J_{21(\alpha)} + J_{21(\beta)} = -J_{12(\alpha)} + J_{21(\beta)}. \tag{6.9}$$

This is consistent with the first two of Eqs. (6.8).

Three-State Cycle. Figure 1.4(b), unmodified, is an example. For the diagram in Fig. 2.5(a), we have the three partial diagrams in Fig. 2.5(b). The arrows can be introduced in three ways, leading to nine directional diagrams [Fig. 2.5(c)]. Thus we find from Eq. (6.3),

$$p_1 = (\alpha_{32}\alpha_{21} + \alpha_{23}\alpha_{31} + \alpha_{21}\alpha_{31})/\Sigma \tag{6.10}$$

$$p_2 = (\alpha_{12}\alpha_{32} + \alpha_{13}\alpha_{32} + \alpha_{31}\alpha_{12})/\Sigma \tag{6.11}$$

$$p_3 = (\alpha_{12}\alpha_{23} + \alpha_{13}\alpha_{23} + \alpha_{21}\alpha_{13})/\Sigma, \tag{6.12}$$

45

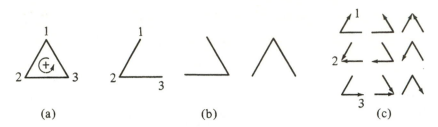

Fig. 2.5. (a) Three-state cycle. (b) Partial diagrams. (c) Directional diagrams.

where Σ is the sum of the three numerators (nine terms), that is, the sum of directional diagrams. The condition for equilibrium is $\Pi_+ = \Pi_-$, or $\alpha_{12}\alpha_{23}\alpha_{31} = \alpha_{21}\alpha_{32}\alpha_{13}$.

Again, because of the single cycle, it is easy to treat the flux (cycle, operational, transition):

$$J = J_{12} = \alpha_{12}p_1 - \alpha_{21}p_2$$

$$= J_{23} = \alpha_{23}p_2 - \alpha_{32}p_3$$

$$= J_{31} = \alpha_{31}p_3 - \alpha_{13}p_1$$

$$\frac{\alpha_{12}\alpha_{23}\alpha_{31} - \alpha_{21}\alpha_{32}\alpha_{13}}{\Sigma} = \frac{\Pi_+ - \Pi_-}{\Sigma}. \tag{6.13}$$

The kinetic differential equations are:

$$\frac{dp_1}{dt} = 0 = J_{21} + J_{31} = -J_{12} + J_{31}$$

$$\tag{6.14}$$

$$\frac{dp_2}{dt} = 0 = J_{12} + J_{32} = J_{12} - J_{23}.$$

Equations (6.14) are consistent with the first three of Eqs. (6.13).

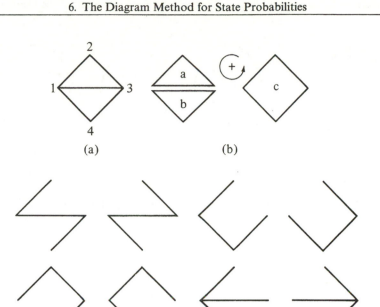

Fig. 2.6. (a) Four-state diagram. (b) Cycles. (c) Partial diagrams.

Simple Three-Cycle Diagram. We begin by pointing out that Fig. 2.6 is a simplification of Fig. 2.1: the two lower states in Fig. 2.1(a) have coalesced to give a diagram, in Fig. 2.6(a), with four states and three cycles. This diagram has 8 partial diagrams, as shown in Fig. 2.6(c); hence the expression for each state probability [Eq. (6.3)] has 8 terms in the numerator (each a product of three rate constants) and 32 terms in Σ, the denominator.

A further condensation and simplification is shown in Fig. 2.7(a): this diagram has two states, three cycles [Fig. 2.7(b)], and three partial diagrams [Fig. 2.7(c)]. From the 6 directional diagrams, we obtain for the steady-state probabilities,

$$p_1 = \frac{\alpha' + \gamma + \beta}{\Sigma}, \quad p_2 = \frac{\alpha + \gamma' + \beta'}{\Sigma} \qquad (6.15)$$

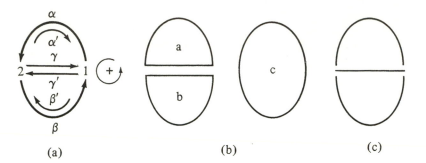

Fig. 2.7. (a) Two-state diagram with three cycles. (b) The cycles. (c) Partial diagrams.

$$\Sigma = \alpha + \alpha' + \beta + \beta' + \gamma + \gamma'.$$

The transition fluxes are

$$J_{12(\alpha)} = \alpha p_1 - \alpha' p_2 = \frac{\alpha\gamma - \alpha'\gamma' + \alpha\beta - \alpha'\beta'}{\Sigma} \qquad (6.16)$$

$$= J_a + J_c \qquad (6.17)$$

$$J_{21(\gamma)} = \gamma p_2 - \gamma' p_1 = \frac{\alpha\gamma - \alpha'\gamma' + \gamma\beta' - \gamma'\beta}{\Sigma} \qquad (6.18)$$

$$= J_a - J_b \qquad (6.19)$$

$$J_{21(\beta)} = \beta p_2 - \beta' p_1 = \frac{\gamma'\beta - \gamma\beta' + \alpha\beta - \alpha'\beta'}{\Sigma} \qquad (6.20)$$

$$= J_b + J_c. \qquad (6.21)$$

Equation (6.17) follows because both cycles a and c use the line 12(α) in the diagram, Fig. 2.7(a) [compare Eqs. (3.3)], and similarly for Eqs. (6.19) and (6.21). Only two of the three transition fluxes in Eqs.

48

(6.16), (6.18), and (6.20) are independent, however, because

$$\frac{dp_1}{dt} = 0 = J_{21(\alpha)} + J_{21(\gamma)} + J_{21(\beta)}$$

$$= -J_{12(\alpha)} + J_{21(\gamma)} + J_{21(\beta)}. \qquad (6.22)$$

Consequently, it is not possible to deduce the three separate cycle fluxes from the transition fluxes. The cycle fluxes can, however, be found from the extension of the diagram method in Section 7. One might guess, from Eqs. (6.16), (6.18), and (6.20), that

$$J_a = \frac{\alpha\gamma - \alpha'\gamma'}{\Sigma}, \quad J_b = \frac{\gamma'\beta - \gamma\beta'}{\Sigma}, \quad J_c = \frac{\alpha\beta - \alpha'\beta'}{\Sigma}, \qquad (6.23)$$

and this, indeed, turns out to be correct. Note that all three cycle fluxes have the form

$$J_\kappa = \frac{\Pi_{\kappa+} - \Pi_{\kappa-}}{\Sigma}. \qquad (6.24)$$

In general it is not possible to deduce the cycle fluxes from the transition fluxes of a diagram (i.e., from the state probabilities). Cycle fluxes are the most interesting and important steady-state kinetic properties of a diagram, especially in free energy transduction systems. Fortunately, the diagram method for state probabilities [Eq. (6.3)] can be extended [2,3] to allow calculation of cycle fluxes, even one-way cycle fluxes. This is the most important topic in the next section.

7. The Diagram Method for Cycle Fluxes and Related Topics

We have previously introduced and compared three kinds of flux (cycle, operational, transition), especially in Chapter 1. We are now in a position, in any example, to express transition (and hence operational) fluxes in terms of the rate constants of the diagram. This is possible because $J_{ij} = \alpha_{ij}p_i - \alpha_{ji}p_j$, and Eq. (6.3) tells us how to find p_i and p_j. Furthermore, we know how to relate a cycle thermodynamic force to the rate constants of the cycle [Eq. (2.27)]. What is missing is the connection between an arbitrary cycle flux J_κ and the rate constants of the diagram that contains the cycle κ. In general, cycle fluxes cannot be deduced from the p_i. The expression of J_κ in terms of diagram rate constants is the first subject addressed in the present section. We then turn to examples and several related topics.

Diagram Method for Cycle Fluxes

We begin by considering net cycle fluxes. This will be followed by a consideration of one-way cycle fluxes. Again, as in Section 6, results will be stated but the proof[2,3] will be omitted.

The method here is analogous to that used for state probabilities p_i in Section 6 but, fortunately, it is simpler to apply. In Section 6, we saw that p_i is the sum of directional diagrams for state i, divided by Σ. Correspondingly, the net flux J_κ in cycle κ of an arbitrary diagram is the sum of flux diagrams belonging to cycle κ, divided by Σ. Flux diagrams are similar to directional diagrams but they are easier to write out because fewer arrows are involved. They are also less numerous. Instead of drawing every possible connected assortment of arrows that flow toward and into a given *state* (as in a directional diagram, Section 6), here we follow the same rules except that the arrows flow into, at one or more states, the *cycle* being considered. A *flux diagram* for a cycle κ is then the cycle itself plus a set of arrows flowing into it. The algebraic value of a flux diagram for cycle κ is the product of rate constants associated with the arrows multiplied by $\Pi_{\kappa+} - \Pi_{\kappa-}$ (the contribution of the cycle). Then the final step is

$$J_\kappa = \text{(sum of all flux diagrams that belong to cycle } \kappa)/\Sigma. \quad (7.1)$$

As an example, consider Fig. 2.8 (an extension of Fig. 1.6). The number of flux diagrams belonging to each of the six cycles is given in Fig. 2.8(b). The 11 possible sets of arrows flowing into cycle a are

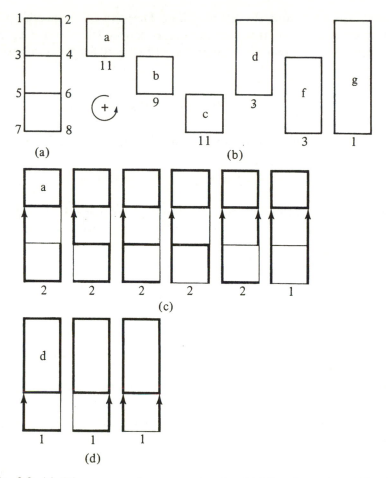

Fig. 2.8. (a) Diagram used as an example. (b) The six cycles with the number of flux diagrams belonging to each cycle. (c) The eleven flux diagrams of cycle a shown explicitly. (d) The three flux diagrams of cycle d shown explicitly.

shown (using symmetry) in Fig. 2.8(c). The sum of the products of these arrows, for an arbitrary cycle κ, is denoted Σ_κ. Hence Σ_a has 11 terms, the first two of which are [see Fig. 2.8(a) for the numbering of states]

$$\Sigma_a = \alpha_{68}\alpha_{87}\alpha_{75}\alpha_{53} + \alpha_{57}\alpha_{78}\alpha_{86}\alpha_{64} + \cdots . \tag{7.2}$$

Each of the 11 terms in Σ_a is multiplied by $\Pi_{a+} - \Pi_{a-}$, that is, by

$$\alpha_{13}\alpha_{34}\alpha_{42}\alpha_{21} - \alpha_{31}\alpha_{43}\alpha_{24}\alpha_{12},$$

to produce the 11 flux diagrams belonging to cycle a. Generalizing, for an arbitrary cycle κ, Eq. (7.1) can be written more explicitly as

$$J_\kappa = \frac{(\Pi_{\kappa+} - \Pi_{\kappa-})\Sigma_\kappa}{\Sigma} = J_{\kappa+} - J_{\kappa-}. \tag{7.3}$$

Thus J_κ is expressed as a function of all the rate constants α_{ij} of the diagram.

Returning to the example in Fig. 2.8, for cycle d [see Fig. 2.8(d)]

$$\Sigma_d = \alpha_{87}\alpha_{75} + \alpha_{78}\alpha_{86} + \alpha_{75}\alpha_{86}. \tag{7.4}$$

Of course Σ_f is similar to Σ_d, by symmetry. Also, because of the symmetry of the diagram,

$$\Sigma_b = \Sigma_d\Sigma_f \text{ (9 terms)}. \tag{7.5}$$

No arrows can flow into cycle g so $\Sigma_g = 1$. Similarly, it should be noted that Eqs. (6.8), (6.13), (6.23), and (6.24) all illustrate cases in which $\Sigma_\kappa = 1$.

Another aspect of the theorem[2,3] on cycle fluxes is the following intuitively plausible relationship: the transition flux J_{ij} between any two neighboring states in the diagram is equal to the sum (with due regard to sign) of cycle fluxes for those cycles that use the line ij. That is, each such cycle makes an additive contribution to J_{ij}. Two examples, from Figs. 2.8(a) and 2.8(b), are

$$J_{35} = J_b + J_d + J_f + J_g$$

$$J_{56} = J_b - J_c + J_d. \qquad (7.6)$$

Simpler examples have already been encountered in Eqs. (6.8), (6.13), (6.17), (6.19), and (6.21). Thus, a given transition flux can be expressed in terms of state probabilities [Eq. (3.1)] or, alternatively, in terms of cycle fluxes. Of course, both modes of expression of the transition flux would lead to the same explicit function of the rate constants α_{ij} of the diagram.

One-Way Cycle Fluxes. Equation (7.3) gives the *net* number of κ cycles completed, in the positive direction, per unit time. One might guess that the two separate one-way cycle fluxes are

$$J_{\kappa+} = \frac{\Pi_{\kappa+}\Sigma_\kappa}{\Sigma} \quad \text{and} \quad J_{\kappa-} = \frac{\Pi_{\kappa-}\Sigma_\kappa}{\Sigma}. \qquad (7.7)$$

This guess is, in fact, correct but it requires proof. "Experimental" evidence[5] for Eqs. (7.7) was found by Monte Carlo simulation in a number of cases. An incorrect proof was given in ref. 3, p. 22. Correct but sophisticated proofs were then found by Kohler and Vollmerhaus[6] and by Qian, Qian, and Qian.[7,8] Finally, a more recent,[9] very simple, proof that Eqs. (7.7) follow from Eq. (7.3) will be given in Section 8.

Both $J_{\kappa+}$ and $J_{\kappa-}$ are always positive but J_κ may be positive or negative. From Eqs. (7.7),

$$\frac{J_{\kappa+}}{J_{\kappa-}} = \frac{\Pi_{\kappa+}}{\Pi_{\kappa-}} = e^{X_\kappa/kT}. \qquad (7.8)$$

Clearly, J_κ and X_κ always have the same sign. This connection between the two one-way cycle fluxes and the cycle force was conjectured in Eq. (3.7) and used in Eq. (5.3). At this point, Eq. (7.8) should be considered to be firmly established.

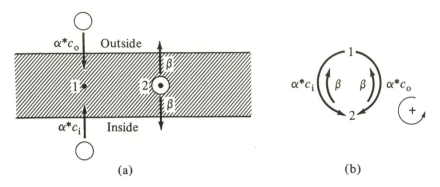

Fig. 2.9. (a) Simple model used to illustrate the nonuniqueness of $J(X)$. (b) The kinetic diagram of the model.

Flux-Force Details in Simple Examples

In this subsection we use three simple examples to illustrate some important details of flux-force relationships that would also be encountered in more complicated models. Here the essential points can be seen with a minimum of algebra.

Nonuniqueness of $J(X)$. Fluxes depend on thermodynamic forces but usually in a nonunique way, except near equilibrium. This is a consequence of the fact that a given force is usually a function of more than one concentration (or other variable) so that the same value of the force can be achieved in more than one way. A very simple model that illustrates this point is shown in Fig. 2.9. A ligand (shown as a circle) can be transported across a symmetrical membrane using single independent binding sites that are either empty (state 1) or occupied by ligand (state 2). The kinetic diagram is given in Fig. 2.9(b). The first-order rate constants are included in the figure. The inside and outside concentrations of ligand are c_i and c_o, respectively. This model is a special case of the two-state model considered in Fig. 2.4 and Eqs. (6.5)–(6.9), but the notation is different.

The thermodynamic force driving ligand from inside to outside is [Eq. (2.2)]

54

$$X = \mu_i - \mu_o = kT \ln(c_i/c_o), \quad e^{X/kT} = c_i/c_o. \tag{7.9}$$

The transport flux, per site, in the same direction, is [Eq. (6.8)]

$$J = \frac{\Pi_+ - \Pi_-}{\Sigma} = \frac{\alpha^* \beta(c_i - c_o)}{\alpha^*(c_i + c_o) + 2\beta}. \tag{7.10}$$

The flux depends on the two separate concentrations but the force depends only on the ratio c_i/c_o. Thus J cannot be a unique function of X, as illustrated explicitly below.

Let the reference equilibrium state be defined by $c_i = c_o = c$, where c is a fixed concentration. At equilibrium, $X = 0$ and $J = 0$. We consider deviations from equilibrium by (a) varying c_i, holding $c_o = c$ and by (b) varying c_o, holding $c_i = c$. In case (a), $c_i = ce^{X/kT}$ and Eq. (7.10) becomes

$$J = \frac{\alpha^* c\beta(e^{X/kT} - 1)}{\alpha^* c(e^{X/kT} + 1) + 2\beta}. \tag{7.11}$$

In case (b), $c_o = ce^{-X/kT}$ and

$$J = \frac{\alpha^* c\beta(e^{X/kT} - 1)}{\alpha^* c(e^{X/kT} + 1) + 2\beta e^{X/kT}}. \tag{7.12}$$

These are two different functions, $J(X)$. However, near equilibrium, $X/kT \to 0$ and both functions yield (see Section 5)

$$J = LX, \quad L = J_{\pm}^e/kT$$

$$J_{\pm}^e = \frac{\alpha^* c\beta}{2(\alpha^* c + \beta)}. \tag{7.13}$$

This illustrates the expression of $J_{\pm}^e$ as an explicit function of rate constants. Near equilibrium,

$$X/kT = (c_i - c_o)/c. \tag{7.14}$$

Flux as a Power Series in the Force.   An equation such as $J = LX$ contains only the first term in a power series in X. To illustrate this, we find J_+, J_-, and J, in a simple model, to quadratic terms in X. The reader may wish to extend the algebra to the terms in X^3.

We use the model in Fig. 2.4 and Eqs. (6.5)–(6.9). We suppose that the force is varied by varying α only (all other rate constants are held fixed). The equilibrium value of α is denoted α^e (also a constant). Then

$$\alpha^e \beta = \alpha' \beta', \quad e^{X/kT} = \frac{\alpha\beta}{\alpha'\beta'} = \frac{\alpha}{\alpha^e}$$

$$\alpha = \alpha^e e^{X/kT} = \alpha^e e^x, \quad x \equiv X/kT.$$

(7.15)

Using Eq. (7.15), we can replace α by the force variable x. Then

$$J_+ = \frac{\alpha\beta}{\alpha + \alpha' + \beta + \beta'} = \frac{\alpha^e \beta e^x}{\alpha^e e^x + \alpha' + \beta + \beta'} = J_- e^x \quad (7.16)$$

$$J_- = \frac{\alpha' \beta'}{\alpha + \alpha' + \beta + \beta'} = \frac{\alpha^e \beta}{\alpha^e e^x + \alpha' + \beta + \beta'} \quad (7.17)$$

$$J = J_+ - J_- = J_-(x)(e^x - 1) = \frac{\alpha^e \beta(e^x - 1)}{\alpha^e e^x + \alpha' + \beta + \beta'}. \quad (7.18)$$

These equations apply at any x. When x is small (near equilibrium), we expand e^x and find

$$J_+ = J_\pm^e \left[1 + (1 - a)x + \left(\frac{1}{2} - \frac{3}{2}a + a^2 \right)x^2 + \cdots \right] \quad (7.19)$$

$$J_- = J_\pm^e \left[1 - ax + \left(-\frac{1}{2}a + a^2 \right)x^2 + \cdots \right] \quad (7.20)$$

$$J = J_+ - J_- = J_\pm^e \left[x + \left(\frac{1}{2} - a \right)x^2 + \cdots \right], \quad (7.21)$$

where

$$J^e_\pm = \frac{\alpha^e \beta}{\alpha^e + \alpha' + \beta + \beta'}, \quad a = \frac{J^e_\pm}{\beta}. \quad (7.22)$$

The coefficients of all powers of x are equilibrium kinetic properties of the model.

Flux-Force Relation by Inspection of Diagram. Any one of Figs. 1.2, 2.1, 2.6, or 2.7 could apply to the M, L transport system first described in relation to Fig. 1.2. Knowledge of the roles played by the three cycles is sufficient to write

$$X_a = X_M, \quad X_b = X_L, \quad X_c = X_M + X_L$$

$$J_M = J_a + J_c, \quad J_L = J_b + J_c. \quad (7.23)$$

The detailed diagram and rate constants are not required here. In fact, one can go further. Define, for the two operational forces,

$$y_M \equiv e^{X_M/kT} - 1, \quad y_L \equiv e^{X_L/kT} - 1. \quad (7.24)$$

Then, for the three cycles, using Eq. (7.8),

$$J_a = J_{a+} - J_{a-} = J_{a-} y_M$$

$$J_b = J_{b+} - J_{b-} = J_{b-} y_L \quad (7.25)$$

$$J_c = J_{c-}[e^{(X_M + X_L)/kT} - 1] = J_{c-}(y_M + y_L + y_M y_L).$$

Thus the operational fluxes become

$$J_M = J_a + J_c = (J_{a-} + J_{c-})y_M + J_{c-} y_L + J_{c-} y_M y_L$$

$$J_L = J_b + J_c = J_{c-} y_M + (J_{b-} + J_{c-})y_L + J_{c-} y_M y_L. \quad (7.26)$$

These equations apply arbitrarily far from equilibrium. Of course forces are involved not only in y_M and y_L but also in J_{a-}, J_{b-}, and J_{c-} (i.e., these are one-way fluxes at the steady state of interest, not at equilibrium). Note the quadratic terms in $y_M y_L$ and also the "reciprocal relation" in the linear terms (the same coefficient J_{c-}^e in the cross terms). These equations make it quite clear that the mere existence of a cycle (c in this case) in which both operational forces are active suffices to introduce (a) coupling between the fluxes and (b) the basis for the flux-force reciprocal relation near equilibrium (Section 5). In fact, from Eqs. (7.26), still without specifying the details of the kinetic diagram, one can write, near equilibrium,

$$J_M = (J_{a\pm}^e + J_{c\pm}^e)(X_M/kT) + J_{c\pm}^e(X_L/kT)$$

$$J_L = J_{c\pm}^e(X_M/kT) + (J_{b\pm}^e + J_{c\pm}^e)(X_L/kT). \tag{7.27}$$

These equations are the same as Eqs. (5.12).

Of course Eqs. (7.26) and (7.27) are formal relations based only on the cycle structure and stoichiometry of the diagram. To make these equations explicit, the diagram must be specified completely (states, rate constants). For example, if we use Fig. 2.7 and Eqs. (6.23) in Eq. (7.26) we have

$$J_{a-} = \frac{\alpha'\gamma'}{\Sigma}, \quad J_{b-} = \frac{\gamma\beta'}{\Sigma}, \quad J_{c-} = \frac{\alpha'\beta'}{\Sigma} \tag{7.28}$$

$$\Sigma = \alpha + \alpha' + \beta + \beta' + \gamma + \gamma'.$$

Number of Independent Transition Fluxes

There is a steady-state transition flux

$$J_{ij} = \alpha_{ij}p_i - \alpha_{ji}p_j = -J_{ji} \tag{7.29}$$

along every line of a diagram. How many of these are independent? For every state of the diagram, we can write a relation between transition fluxes. For example, for Fig. 2.6(a),

$$dp_1/dt = 0 = J_{21} + J_{31} + J_{41}$$

$$dp_2/dt = 0 = J_{12} + J_{32}$$

$$dp_3/dt = 0 = J_{23} + J_{13} + J_{43} \qquad (7.30)$$

$$dp_4/dt = 0 = J_{14} + J_{34}.$$

One of these equations is not independent, however; the sum of the first three equations is the same as the fourth equation. Thus, if the diagram has n states, there are $n - 1$ independent relations of the above type among the J_{ij}. Hence

$$\begin{pmatrix} \text{No. of independent} \\ J_{ij} \end{pmatrix} = \begin{pmatrix} \text{No. of lines} \\ \text{in diagram} \end{pmatrix} - (n - 1). \quad (7.31)$$

For example, in Fig. 2.6(a) there are 5 lines and $n = 4$. Hence the number of independent J_{ij} is 2. Given any two independent J_{ij}, the other three can be deduced from Eqs. (7.30).

In general, knowledge of all the transition fluxes is not sufficient to deduce the cycle fluxes J_κ. For example, Fig. 2.6(a) has only two independent J_{ij} but there are three cycle fluxes: J_a, J_b, and J_c. Two quantities cannot determine three. The state probabilities determine the transition fluxes [Eq. (7.29)] but in general they do not determine the cycle fluxes. The cycle fluxes contain *additional* information. Observation of the state of each system (complex) in a large ensemble of systems at a particular time suffices to determine the state probabilities, but not the cycle fluxes. To find the cycle fluxes, time averaging must be used (see Sections 1 and 8).

A diagram comprised of a single cycle and nothing else is a special case. As the reader can easily verify, all transition fluxes (in the same direction) are equal; hence only one is independent. Also, the transition flux is equal to the cycle flux (same direction).

Transition fluxes are also equal in a portion of a single cycle. Consider, for example, J_{12}, J_{23}, J_{34}, and J_{45} in Fig. 2.10. As in Eqs. (7.30), for states 2, 3, and 4, we have

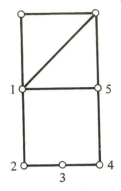

Fig. 2.10. Illustrative diagram.

$$0 = J_{12} + J_{32}$$

$$0 = J_{23} + J_{43} \tag{7.32}$$

$$0 = J_{34} + J_{54}.$$

From these equations we deduce

$$J_{12} = J_{23} = J_{34} = J_{45}. \tag{7.33}$$

This is rather obvious, physically.

Operational fluxes and forces are paired; there is the same number of each. Operational fluxes can be measured, at least in principle, using appropriate transition fluxes. The maximum possible number of independent operational fluxes and forces for a given diagram is equal to the number of independent transition fluxes, as determined by Eq. (7.31). For example, in Fig. 1.6, the diagram has 10 lines and 8 states. Hence, according to Eq. (7.31), the maximum possible number of operational fluxes and forces is 3 (as actually used in the discussion of Fig. 1.6). This number is 1 for a diagram comprised of a single cycle only.

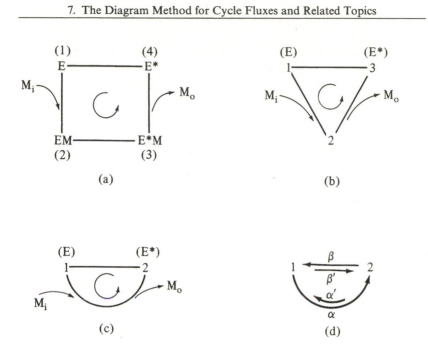

Fig. 2.11. (a) Model for transport of M across a membrane by E. (b) Simplification of the same model. (c) Still further simplification of the same model. (d) Diagram with rate constants for the model in (c).

Two-State Cycle: Stochastics and Tracers

Consider the top cycle (cycle a) in Fig. 1.2(a) as an independent system. This is shown in Fig. 2.11(a). In the positive direction, M is transported across a membrane from an inside bath to an outside bath. Suppose that M is labeled with an isotope in the inside bath only and that both baths are very large. Then a one-way operational flux J_{o+} (in → out) can be obtained by measuring the rate of appearance of the tracer on the outside. In this case the operational or observable process involves the three transitions, 12, 23, and 34, rather than any single transition. Similarly, J_{o-} (out → in) can be found if the label is initially on the outside only.

A simplification is shown in Fig. 2.11(b) where the transitions $EM \rightleftarrows E^*M$ are assumed to be relatively fast. Consequently there is, in effect, a single (fast equilibrium) state 2. Figure 2.11(b) will be considered in Section 8. Here we deal with a still further simplification of the same transport system, the two-state cycle in Fig. 2.11(c). The first-order rate constants are assigned in Fig. 2.11(d). Note that this is the same as Fig. 2.4, some of whose properties are given in Eqs. (6.5)–(6.9). Additional properties, of interest here, are

$$J_+ = \frac{\alpha\beta}{\Sigma}, \quad J_- = \frac{\alpha'\beta'}{\Sigma} \tag{7.34}$$

$$J^{(\alpha)}_{1\to2} = \alpha p_1 = \frac{\alpha\beta + \alpha\alpha'}{\Sigma} > J_+ \tag{7.35}$$

$$J^{(\alpha)}_{2\to1} = \alpha' p_2 = \frac{\alpha'\beta' + \alpha\alpha'}{\Sigma} > J_- \tag{7.36}$$

$$J^{(\beta)}_{2\to1} = \beta p_2 = \frac{\alpha\beta + \beta\beta'}{\Sigma} \neq J^{(\alpha)}_{1\to2}. \tag{7.37}$$

$J^{(\alpha)}_{1\to2}$ is the one-way α transition flux, etc. Even this simple example shows us that some two-way flux relationships do not carry over into the one-way analogues. For example, from Eq. (6.8),

$$J = J_{12(\alpha)} = J_{21(\beta)} = J_o = J_{o+} - J_{o-} \tag{7.38}$$

for the two-way fluxes, but for the corresponding one-way fluxes,

$$J_+ \neq J^{(\alpha)}_{1\to2}, \quad J_+ \neq J^{(\beta)}_{2\to1}$$
$$J^{(\alpha)}_{1\to2} \neq J^{(\beta)}_{2\to1}. \tag{7.39}$$

To see the physical significance of the separate terms in Eqs. (7.35)–(7.37), we turn to a simple stochastic argument. Suppose an ensemble of systems start in state 1 (the same results are found if state 2 is the starting point) and each system undergoes two transitions, the first to state 2 and the second back to state 1. The mean time required for the two transitions [Eq. (1.3)] is

$$\bar{t} = \frac{1}{\alpha + \beta'} + \frac{1}{\alpha' + \beta} = \frac{\Sigma}{(\alpha + \beta')(\alpha' + \beta)}. \tag{7.40}$$

There are four possible outcomes for the two transitions: $\alpha\beta$ ($+$ cycle completed), $\beta'\alpha'$ ($-$ cycle completed), $\alpha\alpha'$ (retrace α transition), and $\beta'\beta$ (retrace β' transition). The respective probabilities are

$$p_+ = \frac{\alpha}{\alpha + \beta'} \cdot \frac{\beta}{\alpha' + \beta} = \frac{\alpha\beta}{(\alpha + \beta')(\alpha' + \beta)} \tag{7.41}$$

$$p_- = \frac{\beta'}{\alpha + \beta'} \cdot \frac{\alpha'}{\alpha' + \beta} = \frac{\alpha'\beta'}{(\alpha + \beta')(\alpha' + \beta)} \tag{7.42}$$

$$p_{\alpha\alpha'} = \frac{\alpha\alpha'}{(\alpha + \beta')(\alpha' + \beta)}, \quad p_{\beta'\beta} = \frac{\beta\beta'}{(\alpha + \beta')(\alpha' + \beta)}. \tag{7.43}$$

The mean rate of completing $+$ cycles is then

$$J_+ = p_+/\bar{t} = \alpha\beta/\Sigma, \tag{7.44}$$

and similarly for the other three possibilities. Thus, for example, the term $\alpha\alpha'/\Sigma$ in Eq. (7.35) is the mean rate at which α, α' sequences occur.

If there is an isotopic tracer on the inside only, the rate at which the tracer appears on the outside is J_{o+}. There are obviously two sequences that contribute to J_{o+}, $\alpha\beta$ and $\alpha\alpha'$. The first sequence corresponds to a $+$ cycle completion in which a labeled M moves from in to out and the second sequence represents an exchange, on the outside, of a labeled molecule (in $\rightarrow$ out) for an unlabeled molecule (out $\rightarrow$ in). That is,

$$J_{ex} = \alpha\alpha'/\Sigma, \quad J_{o+} = J_+ + J_{ex}, \tag{7.45}$$

where J_{ex} is the exchange rate. Note that, in this case,

$$J_{o+} = J^{(\alpha)}_{1 \to 2} = \alpha p_1. \tag{7.46}$$

However, this kind of equality does *not* hold for Fig. 2.11(a) or 2.11(b) (with more than two states in the cycle), as will be seen in Section 8. That is, as already mentioned in the introduction to this subsection, in general a one-way transition flux is *not* an operational (observable) flux.

If the label is initially on the outside, J_{ex} is the same and

$$J_{o-} = J_- + J_{ex} = J^{(\alpha)}_{2 \to 1}. \tag{7.47}$$

Stochastics and a Numerical Example

As explained in Section 1, the sequence of transitions that a complex undergoes may be viewed as a random walk from state to state on the kinetic diagram. We discuss a few more details of such a walk here, and then illustrate some of the points made with a numerical example.

We consider an arbitrary diagram with states i (or j) and with cycles κ. The one-way cycles are $\kappa\pm$, but to simplify notation we use the index η (or η') for one-way cycles instead of $\kappa\pm$. In a very long random walk, governed by the rate constants α_{ij} of the diagram, the fraction of time spent in state i is p_i, the fraction of transitions that start from state i (i.e., the fraction of all state visits that are made to state i) is denoted f_i, and the rate or frequency of η cycle completions is J_η. There is a precise way to define completions[3,5] but this will be omitted here; however, Eq. (1.1) gives some examples.

The total rate of cycle completions is the sum of all the J_η. The reciprocal of this sum is the mean time τ between cycle completions of any type,

$$\tau = \frac{1}{\sum\limits_{\eta} J_\eta}. \tag{7.48}$$

The fraction q_η of all completed cycles that are of type η is

$$q_\eta = \frac{J_\eta}{\sum\limits_{\eta'} J_{\eta'}} = \tau J_\eta. \tag{7.49}$$

Because $J_\eta = \Pi_\eta \Sigma_\eta / \Sigma$ [Eq. (7.7)], we have the explicit expressions

$$\tau = \frac{\Sigma}{\sum_\eta \Pi_\eta \Sigma_\eta} \tag{7.50}$$

and

$$q_\eta = \frac{\Pi_\eta \Sigma_\eta}{\sum_{\eta'} \Pi_{\eta'} \Sigma_{\eta'}}. \tag{7.51}$$

Equation (7.51) for cycles bears some resemblance to Eq. (6.3) for state probabilities. Equations (7.50) and (7.51) will be illustrated below.

Let α_i be the sum of all rate constants *out of* state i. For example, $\alpha_1 = \alpha_{12} + \alpha_{13}$ in Eqs. (1.2) and (1.3). Then, as in Eq. (1.3), the mean time spent in state i, on each visit to this state, is $\bar{t}_i = 1/\alpha_i$. During the course of a long random walk on the diagram, the amount of time spent in state i would be proportional to f_i and also to $\bar{t}_i$. That is, $p_i \propto f_i \bar{t}_i$. On normalizing,

$$p_i = \frac{f_i \bar{t}_i}{\sum_j f_j \bar{t}_j} = \frac{f_i / \alpha_i}{\sum_j (f_j / \alpha_j)}. \tag{7.52}$$

The sum $\sum_i f_i \bar{t}_i$ obviously has the physical significance of the mean time between transitions during the entire walk. We denote this by τ_{tr}. Then, from Eq. (7.52),

$$f_i = p_i \alpha_i \tau_{tr}. \tag{7.53}$$

On summing both sides over i, we find a second expression for τ_{tr}:

$$\tau_{tr} = \sum_i f_i \bar{t}_i = \frac{1}{\sum_i p_i \alpha_i}. \tag{7.54}$$

There is an interesting contrast in the two averages for τ_{tr}.

In numerical calculations the p_i are obtained from the diagram method (Section 6) or by matrix inversion. Then the f_i can be deduced from the p_i and the α_i using Eqs. (7.53) and (7.54):

$$f_i = \frac{p_i \alpha_i}{\sum_j p_j \alpha_j}. \tag{7.55}$$

This equation is used in the following numerical examples.

We adopt Fig. 2.12(a) as an explicit pedagogical example.[3,5] This is a simplified version of Fig. 2.8. The rate constants in Fig. 2.12(a) are in arbitrary units; the cycles are shown in Fig. 2.12(b). The detailed application of the diagram method to this model[3,5] is left as an exercise for the interested reader; we give results only.

Table 2.1 relates to state properties. The second column gives the sum of the directional diagrams (d.d.) for each state [Eq. (6.3)]. The sum of the entries in this column is $\Sigma = 4641$. Then $p_1 = 828/4641$, etc., in the third column. The f_i in the last column follow from Eq.

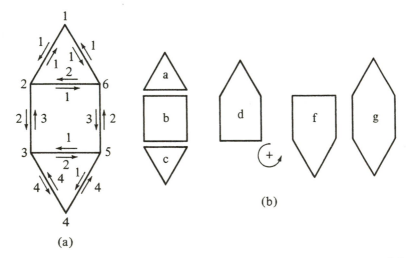

(a)

(b)

Fig. 2.12. (a) Diagram used as a numerical illustration. (b) Cycles of the diagram.

Table 2.1. Properties of States.

State	d.d.i	p_i	α_i	f_i
1	828	0.17841	2	0.07316
2	956	0.20599	4	0.16893
3	532	0.11463	9	0.21152
4	417	0.08985	8	0.14737
5	1208	0.26029	4	0.21347
6	700	0.15083	6	0.18555
Sum	4641	1.00000		1.00000

Table 2.2. Properties of Cycles.

Cycle, η	Π_η	Σ_η	$\Pi_\eta \Sigma_\eta$	$100 J_\eta$
a+	1	148	148	3.1890
a−	2	148	296	6.3779
b+	16	16	256	5.5161
b−	9	16	144	3.1028
c+	16	31	496	10.6874
c−	8	31	248	5.3437
d+	8	8	64	1.3790
d−	9	8	72	1.5514
f+	128	2	256	5.5161
f−	36	2	72	1.5514
g+	64	1	64	1.3790
g−	36	1	36	0.7757
Sum			2152	46.3694

(7.55). The reciprocal of the sum $\Sigma_i p_i \alpha_i$ gives $\tau_{tr} = 0.20503$, the mean time between transitions.

Table 2.2 presents, for each cycle type, Π_η, Σ_η, $\Pi_\eta \Sigma_\eta$, and $100 J_\eta$, where $J_\eta = \Pi_\eta \Sigma_\eta / \Sigma$, with $\Sigma = 4641$ (above). The q_η (not included) follow from the fourth column. For example, $q_{a+} = 148/2152$. The sum of the J_η is 0.4637. The reciprocal of this sum is $\tau = 2.1566$, the mean time between cycle completions. Alternatively, from Eq. (7.50),

$\tau = 4641/2152$. Transitions are of course much more frequent than cycle completions. The mean number of transitions per cycle completion in this example is $\tau/\tau_{tr} = 10.519$.

The above numerical results have been confirmed (to within small fluctuations) by a long Monte Carlo simulation[3,5] of the random walk on Fig. 2.12(a).

8. Recent Advances Concerning Fluxes, Diagrams, and Random Walks

The material in this section is hardly more complicated than in Sections 6 and 7, but it *is* of a supplementary nature. This section is not required for Chapter 3. Five topics are considered.

Mean First Passage Time to Absorption

Random walks with discrete states and continuous time are well-known systems in the stochastic literature. Here we are interested in a special class of such systems in which the walk starts at some specified state and eventually ends at an "absorption" state (each absorption state has a one-way transition into it). In Fig. 2.13(a), the walk starts at state 1 and ends with absorption at state 5 (the rate constants are the usual α_{ij}, with $\alpha_{53} = 0$). In Fig. 2.14(a), the walk starts at state 3 and ends with absorption at either state 1 or state 5. We shall use these two simple examples to illustrate a new

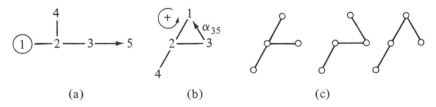

(a)   (b)   (c)

Fig. 2.13. (a) Diagram for a random walk that starts from state 1 (circle) and eventually ends with absorption at state 5. (b) Closed diagram. (c) Partial diagrams belonging to the closed diagram.

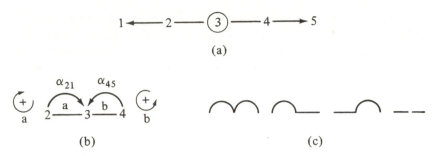

Fig. 2.14. (a) Diagram for a random walk that starts from state 3 (circle) and ends with absorption at either state 1 or state 5. (b) Closed diagram. Cycles a and b are indicated. (c) Partial diagrams of the closed diagram.

and simple way[9] to find the mean time $\bar{t}$ required for absorption to occur (often called the mean first passage time).

The new method has a physical origin or interpretation: when the walk ends at an absorption state, a new walk is begun *immediately* from the same starting state, and this procedure is repeated over and over for a very long time. The mean time to absorption would then be the reciprocal of the rate at which new walks are started during this long time period. But this rate can be found as a steady-state one-way transition flux (or one-way cycle flux), or a sum of such fluxes, in a "closed" diagram in which the absorption transition or transitions are redirected to the starting state (with elimination of the absorption state or states from the diagram). Thus the closed diagram has a new one-way cycle or cycles created by the redirected absorption transition or transitions.

Figure 2.13(b) is the closed diagram derived from the original diagram in Fig. 2.13(a). Note that state 5 is omitted from the closed diagram and that the absorption transition with rate constant α_{35} converts state 3 into state 1 (i.e., immediately restarts the random walk). The other rate constants, α_{ij}, are the same in Figs. 2.13(a) and 2.13(b). Thus a long continuous random walk on the closed diagram simulates *repeated* walks (with immediate restarts) on the original absorption diagram.

69

Similarly, Fig. 2.14(b) is the closed diagram[9] derived from Fig. 2.14(a). Here there are two absorption transitions redirected to the starting state, with elimination of states 1 and 5. The reciprocal of the total rate of arrivals at state 3 in the closed diagram would be the mean time to absorption (see below).

Let us now calculate the mean time to absorption $\bar{t}$ for Fig. 2.13(a), using the closed diagram, Fig. 2.13(b), as described above. The steady-state rate of arrivals at state 1 in the closed diagram can be expressed as a one-way transition flux or as a one-way cycle flux [Eq. (7.7)]:

$$J = \alpha_{35} P_3 = \frac{\Pi_+ \alpha_{42}}{\Sigma} = \frac{\alpha_{12} \alpha_{23} \alpha_{35} \alpha_{42}}{\Sigma}. \tag{8.1}$$

We use P_i for state probabilities in closed diagrams and reserve p_i for original diagrams. To find Σ, we need all the P_i. Figure 2.13(c) shows the partial diagrams belonging to the closed diagram. From these we find [Eq. (6.3)]

$$P_1 = \alpha_{42}(\alpha_{32}\alpha_{21} + \alpha_{23}\alpha_{35} + \alpha_{21}\alpha_{35})/\Sigma$$

$$P_2 = \alpha_{42}\alpha_{12}(\alpha_{32} + \alpha_{35})/\Sigma$$

$$P_3 = \alpha_{42}\alpha_{12}\alpha_{23}/\Sigma \tag{8.2}$$

$$P_4 = \alpha_{24}\alpha_{12}(\alpha_{32} + \alpha_{35})/\Sigma,$$

where Σ is the sum of the four numerators in Eqs. (8.2). It will be seen that $\alpha_{35} P_3$ agrees with the last expression in Eq. (8.1). The mean time to absorption is then

$$\bar{t} = \frac{1}{J} = \frac{\Sigma}{\alpha_{12}\alpha_{23}\alpha_{35}\alpha_{42}}. \tag{8.3}$$

We now make the same kind of calculation for Fig. 2.14(a), using the closed diagram, Fig. 2.14(b). There are two one-way cycles in Fig. 2.14(b), designated a and b.

The state probabilities follow from the partial diagrams in Fig. 2.14(c), or they can easily be found directly from the closed diagram after noting that the total rate constant for $2 \rightarrow 3$ is $\alpha_{23} + \alpha_{21} \equiv B$ and the total rate constant for $4 \rightarrow 3$ is $\alpha_{43} + \alpha_{45} \equiv A$. Thus

$$P_2 = \alpha_{32} A / \Sigma, \quad P_3 = AB / \Sigma, \quad P_4 = \alpha_{34} B / \Sigma$$

$$\Sigma = \alpha_{32} A + \alpha_{34} B + AB. \tag{8.4}$$

Then

$$J_a = \alpha_{21} P_2 = \frac{\Pi_{a+} \Sigma_a}{\Sigma} = \frac{\alpha_{32} \alpha_{21} A}{\Sigma} \tag{8.5}$$

$$J_b = \alpha_{45} P_4 = \frac{\Pi_{b+} \Sigma_b}{\Sigma} = \frac{\alpha_{34} \alpha_{45} B}{\Sigma} \tag{8.6}$$

$$\bar{t} = \frac{1}{J_a + J_b} = \frac{\Sigma}{\alpha_{32} \alpha_{21} A + \alpha_{34} \alpha_{45} B}. \tag{8.7}$$

The probability that absorption occurs at state 1 (rather than at state 5) is

$$\frac{J_a}{J_a + J_b} = \frac{\alpha_{32} \alpha_{21} A}{\alpha_{32} \alpha_{21} A + \alpha_{34} \alpha_{45} B}. \tag{8.8}$$

One-Way Cycle Fluxes from an Expanded Diagram

In complicated kinetic diagrams [e.g., Fig. 2.8(a) is on the verge of being "complicated"], the diagram method becomes impractical: the enumeration of directional and flux diagrams is too tedious. In such a case the p_i of the diagram are easily found numerically and exactly from the steady-state kinetic equations by matrix inversion. However, the J_κ and $J_{\kappa \pm}$ of the diagram cannot be deduced from the p_i; they can be found, approximately, by Monte Carlo simulation of the random walk on the diagram.[3,5] A new method[9] is described here that allows the exact calculation of the $J_{\kappa\pm}$, and hence the J_{κ}, by matrix inversion: Monte Carlo simulation can be avoided. An incidental benefit of the new method is that it provides a transparent proof of the one-way cycle flux relations, Eqs. (7.7).

The new procedure will be illustrated using two simple examples that can easily be handled by the conventional diagram method, but these examples have the virtue of avoiding undue complexity.

The general idea is to choose an arbitrary starting state for the random walk and then construct an expanded diagram that has the same transitions as the original diagram but that includes alternate transition choices up to the completion of the first cycle. All cycles in the original diagram are transformed into one-way cycles in the expanded diagram. The state probabilities in the expanded diagram are designated P_i. Because the expanded diagram contains one-way cycles only, the one-way cycle fluxes can all be found from the P_i and one-way transition fluxes of the expanded diagram. But the P_i can be found by matrix inversion, so this is all that is required to deduce the one-way cycle fluxes.

The first example is the three-state cycle in Fig. 2.15(a). This is the same as Fig. 2.5(a). The state probabilities (of the original diagram) are given in Eqs. (6.10)–(6.12). Let Σ_i be the numerator in the expression for p_i. That is, $p_i = \Sigma_i/\Sigma$. Figure 2.15(b) is the expanded diagram if the random walk starts at state 1 (circled). The two possible transitions out of state 1 are to state 21 or to state 32 (the first index is the actual state; the second index is used to distinguish among repetitions of the same state). Possible transitions out of 21 are to 31 or back to 1. Possible transitions out of 31 are back to 21 or to 1. However, the latter transition would complete a + cycle, as indicated in Fig. 2.15(b). Similar comments can be made about states 32 and 22. The rate constants in Fig. 2.15(b) (use the first index only) are the same as in Fig. 2.15(a) (i.e., the α_{ij}). It will be seen that the original diagram and the expanded diagram represent exactly the same random walk but that the latter diagram contains more detail, namely, how the walk commences out of the starting state. The one-way fluxes in the expanded diagram are

$$J_+ = \alpha_{31}P_{31}, \quad J_- = \alpha_{21}P_{22}. \tag{8.9}$$

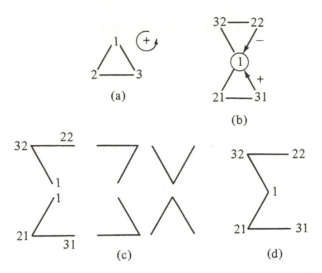

Fig. 2.15. (a) Three-state cycle. (b) Expanded diagram for starting state 1.
(c) Components of the nine partial diagrams that belong to the expanded
diagram. (d) One of the nine partial diagrams.

In the original diagram,

$$J_+ = \frac{\alpha_{12}\alpha_{23}\alpha_{31}}{\Sigma}, \quad J_- = \frac{\alpha_{21}\alpha_{32}\alpha_{13}}{\Sigma}. \tag{8.10}$$

Let us confirm that Eqs. (8.9) and (8.10) agree with each other.

We apply the usual diagram method to the expanded diagram,
Fig. 2.15(b). There are 9 partial diagrams, obtained by combining
any one of the three top pictures in Fig. 2.15(c) with any one of the
three bottom pictures. Figure 2.15(d) gives an example. Then, from
Eq. (6.3),

$$P_1 = \Sigma_1^2/\Sigma', \quad P_{21} = (\alpha_{12}\alpha_{32} + \alpha_{31}\alpha_{12})\Sigma_1/\Sigma'$$

$$P_{31} = \alpha_{12}\alpha_{23}\Sigma_1/\Sigma', \quad P_{22} = \alpha_{13}\alpha_{32}\Sigma_1/\Sigma' \tag{8.11}$$

$$P_{32} = (\alpha_{13}\alpha_{23} + \alpha_{21}\alpha_{13})\Sigma_1/\Sigma'.$$

73

As usual, Σ' (the sum of all directional diagrams) is the sum of the five numerators in Eqs. (8.11):

$$\Sigma' = \Sigma_1(\Sigma_1 + \Sigma_2 + \Sigma_3) = \Sigma_1\Sigma. \qquad (8.12)$$

Thus Σ_1 can be cancelled in each of Eqs. (8.11). We then see that

$$P_1 = p_1, \quad P_{21} + P_{22} = p_2, \quad P_{31} + P_{32} = p_3. \qquad (8.13)$$

That is, states 2 and 3 in the original diagram are subdivided in the expanded diagram. The two one-way fluxes in the expanded diagram are

$$J_+ = \alpha_{31}P_{31} = \alpha_{31} \cdot \alpha_{12}\alpha_{23}/\Sigma$$
$$J_- = \alpha_{21}P_{22} = \alpha_{21} \cdot \alpha_{13}\alpha_{32}/\Sigma, \qquad (8.14)$$

in agreement with Eqs. (8.10).

The expanded diagram is of no practical value in such a simple case but this example illustrates the self-consistency between original and expanded diagrams and also shows how the flux in the original diagram is split explicitly into one-way fluxes in the expanded diagram.

The second example is Fig. 2.16(a), which is the same as Fig. 2.6(a). The three cycles are given in Fig. 2.16(b) and the expanded diagram (with starting state 2) is shown in Fig. 2.16(c). In this case the expanded diagram (11 states) is complicated enough to preclude easy algebra. As an exercise, the reader should construct the expanded diagram independently. Note that cycles end at three different states (31, 2, 11). Relations between original and expanded diagrams are

$$p_1 = P_{11} + P_{12} + P_{13}$$

$$p_2 = P_2$$

$$p_3 = P_{31} + P_{32} + P_{33} \qquad (8.15)$$

$$p_4 = P_{41} + P_{42} + P_{43} + P_{44}$$

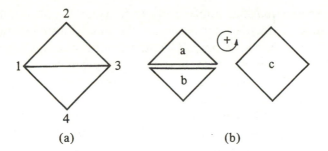

(a)   (b)

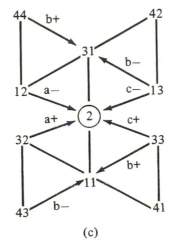

(c)

Fig. 2.16. (a) Diagram used as an example. (b) Cycles. (c) Expanded diagram for starting state 2. The labels on arrows refer to one-way cycle completions.

and

$$J_{a+} = \alpha_{32}P_{32}, \quad J_{a-} = \alpha_{12}P_{12}$$

$$J_{b+} = \alpha_{31}P_{33} + \alpha_{43}P_{44}$$

$$J_{b-} = \alpha_{13}P_{13} + \alpha_{41}P_{43} \tag{8.16}$$

$$J_{c+} = \alpha_{32}P_{33}, \quad J_{c-} = \alpha_{12}P_{13}.$$

75

To use these equations numerically to find the $J_{\kappa\pm}$ one would first calculate the P_i by matrix inversion of the steady-state kinetic equations of the expanded diagram. Of course, in this case it is not difficult to find the $J_{\kappa\pm}$ directly from the original diagram.

Equation (7.3) states that $J_{\kappa+} - J_{\kappa-}$ is proportional to $\Pi_{\kappa+} - \Pi_{\kappa-}$, but is $J_{\kappa\pm}$ proportional to $\Pi_{\kappa\pm}$? The expanded diagram shows that the latter is indeed the case because, when Eq. (7.3) is applied to the expanded diagram (with one-way cycles only), every cycle in the expanded diagram has either $\Pi_{\kappa+} = 0$ or $\Pi_{\kappa-} = 0$. Hence Eqs. (7.7) are established.

Cycle Completions Before Absorption

We return here to the general problem in the first subsection but now consider a special case: Suppose that the original diagram contains one or more cycles and also one or more absorption states. The question then arises[10]: How many cycles of a given type will be completed before absorption occurs? This is a realistic problem in enzyme kinetics. Suppose that an enzyme (or protein complex) is engaged in some kind of cyclic activity. If the enzyme can be degraded at one or more or all of its states by binding a dead-end inhibitor, or by denaturation or proteolysis, any of these events would correspond to "absorption" (termination of the random walk). How much activity does an enzyme molecule or complex accomplish (measured by cycle completions) before it is degraded? We consider two examples.

In Fig. 2.17(a), a random walk starts at state 1 and is terminated, eventually, by absorption at state 5. Except for γ, the rate constants are denoted by the usual α_{ij}. There are six one-way cycles, $\eta = a\pm$, $b\pm$, $c\pm$, as in Figs. 2.6 and 2.16. Every cycle begins and ends at state 1.

The closed diagram (see the first subsection) is Fig. 2.17(b). This diagram has the six cycles (η) just mentioned plus three more cycles involving γ (see below). The total γ (absorption) flux is $J = \gamma P_4$. The mean time to absorption is $\bar{t} = 1/J$. The rate of completing a particular cycle η is J_η. The mean number of cycles η completed in the time $\bar{t}$ is $J_\eta \bar{t} = J_\eta/J$. That is, J_η/J is the mean number of cycles of

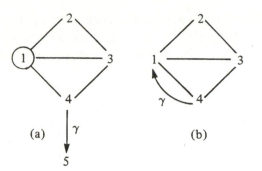

(a)

(b)

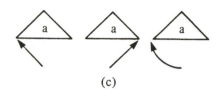

(c)

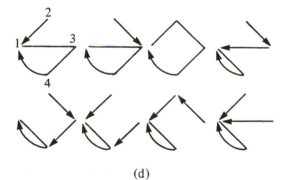

(d)

Fig. 2.17. (a) Diagram for a random walk that starts at state 1 and ends, eventually, with absorption at state 5. (b) Closed diagram. (c) The three flux diagrams for cycle a of the closed diagram. (d) Eight flux diagrams for the three cycles that include the γ transition.

type η completed before absorption occurs for random walks that start at state 1. We apply the diagram method (Sections 6 and 7) to the closed diagram to find J_η and J. Some of the details are left as an exercise for the interested reader.

There are 13 partial diagrams: 8 are shown in Fig. 2.6(c). There are 5 more in which the γ transition replaces the line 1–4. Because the γ transition is one-way, states 1, 2, 3, 4 have 13, 12, 11, 8 directional diagrams, respectively. Hence Σ in Eq. (6.3) is a sum of 44 terms. Of course many terms in Σ include γ, but the γ transition is not involved in the state 4 directional diagrams because it leads *away* from state 4. The six J_η [Eqs. (7.7)] are then

$$J_{a\pm} = \Pi_{a\pm}(\alpha_{41} + \alpha_{43} + \gamma)/\Sigma$$

$$J_{b\pm} = \Pi_{b\pm}(\alpha_{21} + \alpha_{23})/\Sigma \qquad (8.17)$$

$$J_{c\pm} = \Pi_{c\pm}/\Sigma,$$

where $\Pi_{a+} = \alpha_{13}\alpha_{32}\alpha_{21}$, etc. The sum in $J_{a\pm}$ arises from the three flux diagrams in Fig. 2.17(c). The total absorption flux J is [see Fig. 2.17(d)]

$$J = (\text{sum of 8 flux diagrams for } J)/\Sigma = \gamma P_4$$

$$= \gamma(\text{sum of 8 state-4 directional diagrams})/\Sigma. \qquad (8.18)$$

It should be noted that the 8 flux diagrams (all one-way cycles) for J in Fig. 2.17(d) include 3 different cycles and are constructed from the 8 state-4 directional diagrams plus the γ transition.

From Eqs. (8.17) and (8.18), the mean number of completed cycles of type η, before absorption occurs, is then

$$J_\eta \bar{t} = \frac{J_\eta}{J} = \frac{\Pi_\eta \Sigma_\eta}{\gamma(\text{sum of 8 state-4 dir. diag.})}. \qquad (8.19)$$

Note that Σ is not needed for this calculation, though it is for $\bar{t} = 1/J$.

If we define an "event" in the long random walk on the closed diagram as either an absorption or a completion of a cycle of type η, then the probability that an event is an absorption is $J/(J_\eta + J)$ and the probability that it is a cycle η completion is $J_\eta/(J_\eta + J)$. Therefore, the probability that there will be exactly n cycles of type η completed before absorption occurs is obviously

$$\frac{J}{(J_\eta + J)}\left(\frac{J_\eta}{J_\eta + J}\right)^n. \tag{8.20}$$

The mean value of n, from Eq. (8.20), is J_η/J as expected. This simple result, Eq. (8.20), follows because all cycles η, in this example, begin and end at the same state, 1.

The second example is Fig. 2.18. The walk starts at state 1 and absorption occurs from state 3. The two one-way cycles are $\eta = \pm$. The closed diagram is Fig. 2.18(b). The novel feature in this example is that Fig. 2.18(b) may also arise in a quite different context: this is the *original* diagram (not a closed diagram) for a system represented by the square cycle which has, in addition, an irreversible "slip" from state 3 to state 1 with transition probability γ. This slip short-circuits or decouples the business of the square cycle. After the γ transition (slip), the random walk continues from state 1. In the absorption problem, Fig. 2.18(a), the walk on Fig. 2.18(b) is conceptual (we imagine the walk to be restarted, instantaneously, over and over); in the slip problem, the walk on Fig. 2.18(b) is real. The properties of Fig. 2.18(b) are the same in either case.

The numbers of directional diagrams in Fig. 2.18(b) are 8, 6, 4, 6 for states 1, 2, 3, 4, respectively (see Fig. 2.6). Thus Σ has 24 terms. As in Eq. (8.19),

$$\frac{J_\pm}{J} = \frac{\Pi_\pm}{\gamma(\text{sum of 4 state-3 dir. diag.})}. \tag{8.21}$$

The denominator on the right is also the sum of 4 flux diagrams for J (absorption or slip), shown in Fig. 2.18(c). Equation (8.21) gives the mean number of completed $\pm$ cycles before absorption or per slip transition. Equation (8.20) is again applicable.

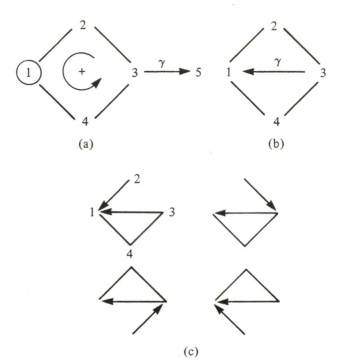

Fig. 2.18. (a) Random walk with starting state 1, absorption state 5. (b) Closed diagram, or the original diagram with a "slip" transition included. (c) Four flux diagrams for the two cycles that include the γ transition.

Tracers and One-Way Operational Flux

This subject was discussed briefly in Section 7 in relation to a two-state cycle, using a stochastic argument. Here we consider a more general method for this kind of problem, included in recent work by Chen.[11] An expanded diagram is used that is different from but reminiscent of the type of expanded diagram already introduced in Figs. 2.15 and 2.16.

We return to the model in Fig. 2.11(a), and suppose first that the ligand M is labeled in the inside bath. We are interested in the steady-state rate of appearance of label in the outside bath (this is J_{o+}, the

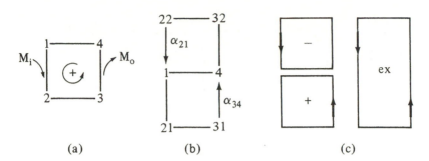

Fig. 2.19. (a) Diagram for transport of M across a membrane. (b) Expanded diagram for tracer calculations. (c) The three cycles of the expanded diagram.

operational flux in the + direction) and in the subdivision of J_{o+} into J_+ (+ cycles) and J_{ex} (exchange in the outside bath, of labeled M for unlabeled M, without completing a + cycle).

Chen's method is indicated in Fig. 2.19 for this particular example. Figure 2.19(a) is the original diagram. The rate constants are the usual α_{ij}. Figure 2.19(b) is the expanded diagram that is used for the calculations. Note the duplication of states 2 and 3, requiring a second index, and also the one-way arrows (transitions). The same rate constants α_{ij} apply to original and expanded diagrams. Also, the same transition choices are available in original and expanded diagrams.

The expanded diagram has three cycles, all of which are one-way cycles [Fig. 2.19(c)]. Chen has shown that: the + cycle flux in the expanded diagram is the same as J_+ in the original diagram; the − cycle flux in the expanded diagram is the same as J_- in the original diagram; and the "ex" cycle flux in the expanded diagram is the exchange rate, J_{ex}. Then

$$J_{o+} = J_+ + J_{ex} = \alpha_{34}P_{31},\qquad(8.22)$$

where P_{31} is the steady-state probability of state 31 in the expanded diagram. The flux $\alpha_{34}P_{31}$ is a transition flux that is comprised of

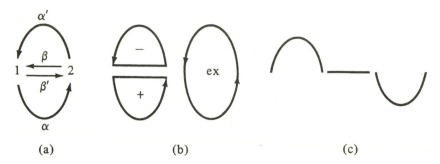

(a)   (b)   (c)

Fig. 2.20. (a) Expanded diagram from Fig. 2.11(d). (b) Cycles of the expanded diagram. (c) Partial diagrams of the expanded diagram.

contributions from two cycles (+ and ex). Note that the sequence of states around the ex cycle corresponds, in the original diagram, to the sequence 1234321. This sequence obviously accomplishes the label exchange.

To simplify algebra, we work out details in this example for the two-state and three-state versions of Fig. 2.19 [see Figs. 2.11(b) and 2.11(c)], but not for the four-state case just discussed.

In the two-state case, the original diagram is Fig. 2.11(d) and the expanded diagram is Fig. 2.20(a). The cycles of the expanded diagram are in Fig. 2.20(b) and the partial diagrams are in Fig. 2.20(c). The state probabilities are denoted P_1 and P_2 and the sum of directional diagrams is Σ'. We find immediately that $\Sigma' = \Sigma$, $P_1 = p_1$, and $P_2 = p_2$ [see Eqs. (6.5)]. The cycle fluxes and J_{o+} for the expanded diagram are (all $\Sigma_\kappa = 1$)

$$J_+ = \frac{\alpha\beta}{\Sigma}, \quad J_- = \frac{\alpha'\beta'}{\Sigma}, \quad J_{ex} = \frac{\alpha\alpha'}{\Sigma}, \quad J_{o+} = \alpha P_1, \qquad (8.23)$$

in agreement with Eqs. (7.34) and (7.45). We see that this example is so simple that Chen's method adds nothing new. However, self-consistency has been tested.

In the somewhat more complicated three-state case, the original diagram is Fig. 2.11(b), which is the same as Fig. 2.5(a). Some of the

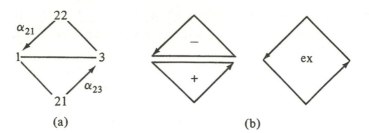

(a)   (b)

Fig. 2.21. (a) Expanded diagram for Fig. 2.11(b). (b) Cycles of the expanded diagram.

properties of this diagram, including the p_i, are given in Eqs. (6.10)–(6.13). The expanded diagram is Fig. 2.21(a), which has the three cycles shown in Fig. 2.21(b). The partial diagrams belonging to the expanded diagram are those in Fig. 2.6(c). The sum of all directional diagrams is denoted Σ'. From the partial diagrams, we find [Eq. (6.3)]

$$P_1 = p_1, \quad P_3 = p_3, \quad P_{21} + P_{22} = p_2$$

$$\Sigma' = (\alpha_{21} + \alpha_{23})\Sigma,$$

(8.24)

where Σ is defined following Eq. (6.12). The cycle fluxes of the expanded diagram are

$$J_+ = \frac{\alpha_{12}\alpha_{23}\alpha_{31}(\alpha_{21} + \alpha_{23})}{\Sigma'} = \frac{\alpha_{12}\alpha_{23}\alpha_{31}}{\Sigma}$$

(8.25)

$$J_- = \frac{\alpha_{13}\alpha_{32}\alpha_{21}(\alpha_{21} + \alpha_{23})}{\Sigma'} = \frac{\alpha_{13}\alpha_{32}\alpha_{21}}{\Sigma}$$

(8.26)

$$J_{ex} = \frac{\alpha_{12}\alpha_{23}\alpha_{32}\alpha_{21}}{\Sigma'}.$$

(8.27)

J_+ and J_- agree with Eq. (6.13).

The one-way operational (tracer) flux J_{o+} is $J_+ + J_{ex}$. In the original diagram, the two relevant one-way transition fluxes are [Eqs. (6.10) and (6.11)]

$$J_{1 \to 2} = \alpha_{12} p_1 \quad \text{and} \quad J_{2 \to 3} = \alpha_{23} p_2. \qquad (8.28)$$

If Σ' is used in the denominator in all of Eqs. (8.25), (8.27), and (8.28), we see that the three terms in the numerator in J_{o+} appear in $J_{1 \to 2}$ and $J_{2 \to 3}$, but each of the two latter fluxes also have three more terms (i.e., both are larger than J_{o+}). Hence, neither one-way transition flux is the operational flux. Also, $J_{1 \to 2} \neq J_{2 \to 3}$.

If the tracer is placed in the outside bath, one finds that J_{ex} is the same as above and that

$$J_{o-} = J_- + J_{ex} \qquad (8.29)$$

$$J_o \equiv J_{o+} - J_{o-} = J_+ - J_- = J$$
$$= J_{12} = J_{23} = J_{31}. \qquad (8.30)$$

Thus the two-way relations are quite different from the one-way relations.

Diagram Method Applied to Discrete-Time Problems

So far in this book all applications of the diagram method have been in continuous time. However, the method can be applied to discrete-time problems as well.[12] In such problems each transition (event) counts as one unit of time. Instead of a rate constant α_{ij} (probability per unit time—see Section 1) for a transition $i \to j$, we have here a probability α_{ij}. Furthermore, the sum of all possible probabilities (for transitions) out of any state i must be unity, i.e., $\Sigma_j \alpha_{ij} = 1$. We have previously called this sum α_i. The time spent in each state before a transition is $1/\alpha_i = 1$.

We consider a single example. In a two-out-of-three set tennis match, player 1 has a probability p of winning each set and player 2 has a probability q of winning, where $p + q = 1$. We are interested in the probability that, say, player 1 wins the match and also in the mean number of sets played per match (if the match were to be repeated many times). The latter quantity is a mean first passage time in an absorption problem: the kinetic diagram, including four absorption states, is shown in Fig. 2.22(a). The state labels give the score in sets. There are one-way transitions only. Each transition

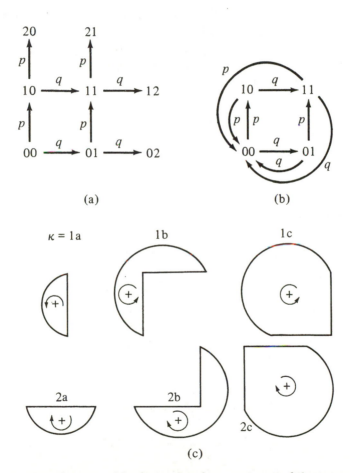

(a)

(b)

(c)

Fig. 2.22. (a) Diagram with absorption for a two-out-of-three set tennis match. (b) Closed diagram. (c) Cycles.

represents a set played. Absorption corresponds to completion of the match. Player 1 wins if state 20 or state 21 is reached.

This problem is easy to solve directly (as in textbooks) but we treat it by the method introduced at the beginning of this section. The closed diagram is shown in Fig. 2.22(b) (the starting state in the original diagram is 00). The closed diagram will give us average properties corresponding to repetition of the match many times.

The closed diagram has 30 partial diagrams, 10 of which make no contribution to any state. Each of the other 20 makes a contribution to only one state. It is much simpler in this problem to find the state probabilities P_{ik} directly from the rather simple steady-state kinetic equations [compare Eqs. (6.1)]:

$$0 = p(P_{10} + P_{11}) + q(P_{01} + P_{11}) - P_{00}$$

$$0 = pP_{00} - P_{10}, \quad 0 = qP_{00} - P_{01} \tag{8.31}$$

$$0 = qP_{10} + pP_{01} - P_{11}.$$

The solution is

$$P_{00} = 1/\Sigma, \quad P_{10} = p/\Sigma, \quad P_{01} = q/\Sigma$$

$$P_{11} = 2pq/\Sigma, \quad \Sigma = 2(1 + pq). \tag{8.32}$$

These are the probabilities of occurrence of the different scores during a long series of repeated matches.

The six cycles (all one-way) are shown and labeled in Fig. 2.22(c). Application of Eq. (7.3) leads to

$$J_{1a} = p^2/\Sigma, \quad J_{1b} = J_{1c} = p^2 q/\Sigma$$

$$J_{2a} = q^2/\Sigma, \quad J_{2b} = J_{2c} = pq^2/\Sigma. \tag{8.33}$$

Transition probabilities confirm these fluxes:

$$pP_{10} = J_{1a}, \quad pP_{11} = J_{1b} + J_{1c}$$

$$qP_{01} = J_{2a}, \quad qP_{11} = J_{2b} + J_{2c}. \tag{8.34}$$

The total fluxes for each player, that is, the number of matches won per set played is

$$J_1 \equiv J_{1a} + J_{1b} + J_{1c} = p^2(1 + 2q)/\Sigma$$

$$J_2 \equiv J_{2a} + J_{2b} + J_{2c} = q^2(1 + 2p)/\Sigma. \tag{8.35}$$

The mean total number of matches completed per set played is

$$J_1 + J_2 = 1/\Sigma = P_{00}. \tag{8.36}$$

This equation states that the total flow into state 00 ($J_1 + J_2$) is equal to the total flow out of state 00 (i.e., P_{00}). The mean time to absorption is the mean number of sets per match:

$$\bar{t} = 1/(J_1 + J_2) = \Sigma = 2(1 + pq). \tag{8.37}$$

The probability that player 1 wins any match (in the long series of matches) is

$$\frac{J_1}{J_1 + J_2} = p^2(1 + 2q). \tag{8.38}$$

For player 2,

$$\frac{J_2}{J_1 + J_2} = q^2(1 + 2p). \tag{8.39}$$

It is also possible, using the methods of this section, to calculate the number of visits to a given state before absorption. But this subject[13] gets a little complicated, so it is omitted.

References

1. King, E. L. and Altman, C. (1956) J. Phys. Chem. **60**, 1375.
2. Hill, T. L. (1966) J. Theoret. Biol. **10**, 442.
3. Hill, T. L. (1977) *Free Energy Transduction in Biology* (Academic, New York).
4. Caplan, S. R. and Zeilberger, D. (1982) Adv. Applied Math. **3**, 377.
5. Hill, T. L. and Chen, Y. (1975) Proc. Natl. Acad. Sci. USA **72**, 1291.
6. Kohler, H.-H. and Vollmerhaus, E. (1980) J. Math. Biol. **9**, 275.
7. Qian, C., Qian, M. and Qian, M. (1981) Scientia Sinica **24**, 1431.
8. Qian, M., Qian, M. and Qian, C. (1984) Scientia Sinica **27**, 470.
9. Hill, T. L. (1988) Proc. Natl. Acad. Sci. USA **85**, 2879.
10. Hill, T. L. (1988) Proc. Natl. Acad. Sci. USA **85**, 3271.
11. Chen, Y., to be published.
12. Hill, T. L. (1988) Proc. Natl. Acad. Sci. USA **85**, 5345.
13. Hill, T. L. (1988) Proc. Natl. Acad. Sci. USA **85**, 4577.

3

Free Energy Levels and Application to Muscle Contraction

A supplementary topic in relation to free energy transduction (Chapter 1) and the diagrams we have been discussing (Chapter 2) is the assignment of relative free energy levels to the states of the diagram. For most kinetic systems, this is of conceptual value, primarily, except for a new expression for the rate of free energy dissipation (introduced in Sections 4 and 5). However, free energy levels are much more important in understanding muscle contraction and related systems (dynein, kinesin).

We give a brief account in this chapter of free energy levels for ordinary kinetic diagrams (Section 9) and their role in muscle contraction (Section 10). Many more details are available elsewhere.[1-6]

9. Free Energy Levels in Kinetic Diagrams[1]

The enzyme or protein complex of interest can exist in a finite number of discrete states, for example, in Fig. 1.2, E, E*, EM, etc. As usual, we use the index i for these states. The fraction of all complexes in state i in an ensemble of complexes is p_i; p_i may be a function of time before steady state is reached. The chemical potential of a complex in state i is written

$$\mu_i = G_i + kT \ln p_i, \qquad (9.1)$$

where G_i is a standard chemical potential with standard state $p_i = 1$ (all complexes in state i). That is, G_i is the chemical potential (Gibbs free energy per complex) of complexes in state i when all complexes are in state i.

Whereas G_i in Eq. (9.1) is an intrinsic property of state i alone, the chemical potential μ_i is obviously a property of the whole ensemble of complexes since the population p_i at an arbitrary time t depends in general on *all* the rate constants of the diagram and on the initial state—at $t = 0$—of the ensemble and on any restraints imposed in the time evolution of the ensemble.

The standard free energies G_i of the states i of a single complex are fundamental in that they can be related to the first-order rate constants (of the diagram) which determine the kinetics of the ensemble. To see this, we have to examine two cases. Consider first the inverse transitions between two states i and j such that no ligand is bound or released in the transitions [e.g., a conformational change $E \rightleftharpoons E^*$ or $ES \rightleftharpoons EP$ in Fig. 1.4(b)]. That is, these are *isomeric* transitions. As usual, the first-order rate constants for $i \rightarrow j$ and $j \rightarrow i$ are designated α_{ij} and α_{ji}, respectively. A hypothetical equilibrium between the two states (with all other transitions in the diagram imagined blocked) can be used to establish the connection between the G's and α's:

$$\alpha_{ij} p_i^e = \alpha_{ji} p_j^e \quad \text{(detailed balance)} \qquad (9.2)$$

$$\mu_i = G_i + kT \ln p_i^e = \mu_j = G_j + kT \ln p_j^e \qquad (9.3)$$

$$\alpha_{ij}/\alpha_{ji} = \exp[-(G_j - G_i)/kT] \equiv K_{ij}, \qquad (9.4)$$

where K_{ij}, defined here, is a dimensionless equilibrium constant. Although $\alpha_{ij}/\alpha_{ji} = p_j^e/p_i^e$ holds only at equilibrium, Eq. (9.4) is valid regardless of the state of the ensemble, that is, even after the imaginary blocks on other transitions of the diagram are removed. This follows because the G's and α's are *intrinsic properties* of states i and j of *each complex* that can have nothing to do with what is transpiring in the rest of the ensemble. We have merely used the above hypothetical equilibrium condition of the ensemble, as a convenience, to establish a relation that is *independent* of the state of the ensemble.

The second possibility we have to examine is that a ligand M (or S, or P, etc.) is bound in the transition $i \to j$. In this case, if we consider the equilibrium between states i and j in the presence of M in solution at its *actual* concentration c_M, we have

$$\alpha_{ij}^* c_M p_i^e = \alpha_{ji} p_j^e \quad \text{(detailed balance)}$$

or

$$\alpha_{ij} p_i^e = \alpha_{ji} p_j^e, \tag{9.5}$$

where the (pseudo) first-order rate constant $\alpha_{ij} \equiv \alpha_{ij}^* c_M$. The equilibrium condition is now

$$\mu_i + \mu_M(c_M) = G_i + kT \ln p_i^e + \mu_M = \mu_j = G_j + kT \ln p_j^e, \tag{9.6}$$

where μ_M is given in Eq. (2.1). Therefore

$$\frac{\alpha_{ij}}{\alpha_{ji}} = \exp \left\{ -\frac{[G_j - (G_i + \mu_M)]}{kT} \right\} \equiv K_{ij}, \tag{9.7}$$

where again K_{ij}, defined here, is a dimensionless equilibrium constant. K_{ij} depends on the value of c_M. Thus, in binding transitions, μ_M (or μ_S, etc.) must be included in the free energy difference to establish the correct relationship to the first-order rate constant ratio α_{ij}/α_{ji}. Equation (9.7) is the analogue of Eq. (9.4). We see that by "correcting" G_i with $\mu_M(c_M)$, both types of transition can be put on the same footing: a free energy change between states is related to the quotient of first-order rate constants.

The kind of free energy change between states i and j that appears in Eqs. (9.4) and (9.7), which is directly related to *first-order* rate constants, is referred to as a *basic free energy* change. If a ligand M at c_M is involved in the process, $\mu_M(c_M)$ must be included, as illustrated in Eq. (9.7). The term "basic" is appropriate because: (a) these free energy changes are time-independent intrinsic properties of each individual complex of the ensemble; (b) the use of first-order rate constants throughout the diagram puts all transitions on an equivalent *kinetic* and mathematical basis; (c) the "correction" with μ_M (or μ_S, etc.), as needed, in the basic free energy change correspondingly puts all of the states of the diagram on an equivalent *thermodynamic* basis so that all transitions become pseudoisomeric. The "correction" with μ_M is the thermodynamic equivalent of the "correction" of a second-order rate constant α_{ij}^* to give a pseudo-first-order rate constant $\alpha_{ij} = \alpha_{ij}^* c_M$ (see Section 1). This uniform first-order isomeric formalism is the simplest possible.

Ordinary biochemical thermodynamics focuses attention on free energy changes involving substrates, products, ligands, etc. Here the point of view is just the opposite. In order to relate in the simplest way possible to first-order diagram kinetics, we deal with pseudo-isomeric *macromolecular* states, free energy levels, and transitions in which substrates, etc., are relegated to a strictly secondary and implicit role. Of course the substrates, etc., come to the fore at the complete cycle level because they then account for the only changes that have taken place as a result of a cycle completion.

A second free energy change of importance relates to the ensemble as a whole, at an arbitrary time t, whereas basic free energy changes are "private" properties of individual complexes (and therefore do not depend at all on initial conditions or the time t). If the ensemble at time t has a composition specified by $p_1, p_2, \ldots, p_n$, then the free energy change *in the ensemble* plus surrounding bath or baths, associated with the processes $i \rightarrow j$ in Eqs. (9.4) and (9.7) are, respectively,

$$\mu_j - \mu_i = G_j - G_i + kT \ln(p_j/p_i)$$

$$\mu_j - (\mu_i + \mu_M) = G_j - (G_i + \mu_M) + kT \ln(p_j/p_i).$$

(9.8)

Since these changes refer to the whole ensemble, they are designated *gross free energy* changes.

For convenience, for both kinds of free energy changes, we will refer as well to *free energy levels* of the separate *states*. For example, G_j and G_i in Eq. (9.4) and G_j and $G_i + \mu_M$ in Eq. (9.7) are the "basic free energy levels" of state j and i. Actually, only the differences are significant but the assignment of separate levels to the states has value as a conceptual aid. Note that, in the second example here, we could just as well have used $G_j - \mu_M$ and G_i.

The gross free energy changes in Eqs. (9.8) are, of course, just the conventional Gibbs free energy changes that determine equilibrium in the ensemble (plus baths) or the direction of spontaneous "flow" towards equilibrium. A new name ("gross") is adopted for convenience (to contrast with "basic") but one is also needed because it refers to the relative free energy *levels* of the *macromolecular* states on a *pseudoisomeric* basis. As explained above, this is not a conventional point of view; therefore, some new designation is appropriate.

Although we shall speak, for brevity, of the free energy levels of the *macromolecular* states, it must be kept in mind that the corresponding free energy changes relate to macromolecules *plus* surrounding bath or baths.

From an operational point of view, in theoretical work, basic free energy levels are introduced *ab initio* as fixed parameters of a model, while the gross free energy levels emerge as calculated macroscopic properties of the ensemble that depend on the p_i and therefore on the initial conditions and on the time t, in general. The gross levels are of course all equal when there is (pseudoisomeric) equilibrium among all macromolecular states i, but not otherwise.

In its stochastic behavior, any *individual* complex of the ensemble is governed entirely by the α's of the diagram (which are related to the basic free energy levels). The individual complex has no knowledge of *ensemble* properties such as the p's, the gross free energy levels, transient versus steady state, etc.—or even whether there *is* an ensemble. Ensemble properties depend on the *statistics* of these individually completely uncontrolled stochastic systems. Of course the choice of rate constants, the α's, for the individual complexes will determine whether, at $t = \infty$, the ensemble statistics will correspond to equilibrium or to a nonequilibrium steady state.

It is convenient to introduce special notation for basic and gross free energy changes and levels. For the process $i \to j$, whether it be an isomeric change or it involves binding or release of a ligand, etc., we use

$$\Delta G'_{ij} = G'_i - G'_j, \quad \Delta \mu'_{ij} = \mu'_i - \mu'_j \qquad (9.9)$$

to designate the basic and gross free energy changes, respectively. The quantities on the right represent the basic and gross free energy levels of the states. The primes indicate, on both sides of the equation, that μ_M (or μ_S, etc.) is included *where needed*—i.e., that macromolecular free energies are "corrected," as appropriate, by ligand, substrate, etc., chemical potentials in order to put all states in the diagram on the same (isomeric) footing.

Note that Δ is used in Eqs. (9.9) (and below) in an unconventional way: $\Delta \equiv$ initial $-$ final, rather than vice versa. This is done because most steady-state applications are dominated by downhill (in free energy) transitions. Hence *positive* values of $\Delta G'_{ij}$ and $\Delta \mu'_{ij}$ predominate.

With these definitions of $\Delta G'_{ij}$ and $\Delta \mu'_{ij}$, we now have the important relations, for any change in state $i \to j$,

$$\alpha_{ij}/\alpha_{ji} = \exp(\Delta G'_{ij}/kT) = K_{ij} \qquad (9.10)$$

and

$$\Delta \mu'_{ij}(t) = \Delta G'_{ij} + kT \ln[p_i(t)/p_j(t)]. \qquad (9.11)$$

If states i and j are in equilibrium with each other, $\Delta \mu'^e_{ij} = 0$.

The product of Eq. (9.10) around any cycle κ, say with states numbered $1, 2, \ldots, m$ (in the positive direction), is [Eq. (2.27)]

$$\Pi_{\kappa+}/\Pi_{\kappa-} = e^{X_\kappa/kT} = K_{12}K_{23}\cdots K_{m1}. \qquad (9.12)$$

Also, the total thermodynamic force in the cycle, X_κ, is equal to the sum of the successive basic free energy changes around the cycle:

$$X_\kappa = \Delta G'_{12} + \Delta G'_{23} + \cdots + \Delta G'_{m1}. \qquad (9.13)$$

If X_κ is positive, not all of the $\Delta G'_{ij}$ here need be positive. We also have

$$X_\kappa = \Delta\mu'_{12}(t) + \Delta\mu'_{23}(t) + \cdots + \Delta\mu'_{m1}(t) \qquad (9.14)$$

because the probability terms cancel if Eq. (9.11) is summed around the cycle (at arbitrary t). X_κ is, of course, independent of t because it depends only on fixed concentrations of ligands, substrates, etc.

The overall "drive" around the cycle κ is provided by X_κ. This is both an operational thermodynamic force, or sum of such forces, and a function of the cycle rate constants, as given by Eq. (9.12). This of course implies that, in making up any model, the rate constants of each cycle must be selected to be consistent with the thermodynamic forces in the cycle.

Equation (9.13) shows how the overall drive or force is subdivided among the separate transitions of the cycle according to the $\Delta G'_{ij}$ (or α_{ij}/α_{ji}) values. Thus, $\Delta G'_{ij}$ is the "drive" associated with the transition ij. This subdivision again emphasizes that, in general, no transition or "step" can be singled out as completely controlling the steady-state cyclic action. All steps are involved, one way or another. By the same token, if the cycle contains two or more forces so that free energy transduction is possible, the transduction is an indivisible property of the whole cycle and cannot be assigned to any one transition or $\Delta G'_{ij}$. In any case, the *kinetic* behavior of the ensemble (e.g., at steady state) is by no means completely determined by the $\Delta G'_{ij}$ values.

The subdivision in Eq. (9.13) is an invariant ("basic") property (of any given model) that belongs to each *individual* complex in the ensemble. On the other hand, the subdivision in Eq. (9.14) is a time-dependent property of the ensemble taken as a whole.

Direction of Spontaneous Transition

Consider an ensemble of complexes with an arbitrary kinetic diagram. At an arbitrary time t (this need not be a steady state), let p_i be the probability of state i. For any transition ij in the diagram, the net mean transition flux $i \to j$ is [Eq. (3.1)]

$$J_{ij}(t) = \alpha_{ij} p_i(t) - \alpha_{ji} p_j(t). \tag{9.15}$$

Also, Eqs. (9.10) and (9.11) can be combined to give

$$\alpha_{ij} p_i(t)/\alpha_{ji} p_j(t) = \exp[\Delta\mu'_{ij}(t)/kT]. \tag{9.16}$$

On comparing Eqs. (9.15) and (9.16), we see that the transition flux J_{ij} for any transition ij *always has the same sign* as the gross free energy level difference $\Delta\mu'_{ij}$ (at any time t). For example, if i has the higher gross free energy level, the net mean flux will be in the direction $i \to j$ (there are, of course, stochastic exceptions in single complexes or in small groups of complexes). This is just the second law of thermodynamics at work on individual reactions of the diagram.

Thus, net reaction (positive flux) always occurs in a *downhill* direction with reference to a set of gross free energy levels. This is *not* true of the invariant basic free energy levels.

Rate of Free Energy Dissipation

Each event $i \to j$ in a single complex contributes a free energy drop $\Delta\mu'_{ij}$ [Eqs. (9.8)] to the whole ensemble (plus baths). Hence $J_{ij}\Delta\mu'_{ij}$ is the net rate of free energy dissipation contributed by the process $i \to j$. We have just seen that the product $J_{ij}\Delta\mu'_{ij}$ is always ≥ 0, with the equality holding only at equilibrium. The *total* rate of free energy dissipation in ensemble plus baths, at any time t, is therefore

$$\Phi = \sum_{ij} J_{ij}(t)\Delta\mu'_{ij}(t) \geq 0, \tag{9.17}$$

where the sum is over all lines in the diagram. (The direction chosen along each line is immaterial because a change in direction reverses both signs in the product.) Each term in the sum is never negative.

Equation (9.17) should be compared with Section 4. This equation resembles Φ expressed in terms of cycle fluxes (each term in the sum is never negative) but differs from Φ expressed in terms of operational fluxes (some term or terms may be negative).

Simple Enzyme-Substrate-Product Example

We use Fig. 1.4(b) to illustrate some of the general comments made above. Equations (2.13)–(2.20) should be reviewed.

The equations of type (9.10) for this model are

$$\alpha_{12}/\alpha_{21} = \exp\{[G_1 + \mu_S) - G_2]/kT\} = K_{12}$$

$$\alpha_{23}/\alpha_{32} = \exp[(G_2 - G_3)/kT] = K_{23} \qquad (9.18)$$

$$\alpha_{31}/\alpha_{13} = \exp\{[G_3 - (G_1 + \mu_P)]/kT\} = K_{31}.$$

On multiplying these together we get

$$\Pi_+/\Pi_- = e^{(\mu_S - \mu_P)/kT} = e^{X/kT} = K_{12}K_{23}K_{31}, \qquad (9.19)$$

where μ_S and μ_P are given by Eqs. (2.14). The quantity $\mu_S - \mu_P$ is the actual free energy drop (usually positive) associated with one counterclockwise cycle (one S at c_S is converted to one P at c_P). The equilibrium case is $\mu_S = \mu_P$ [all four members of Eq. (9.19) are equal to unity].

Figure 3.1(a) shows a hypothetical (infinite) set of basic free energy levels for this system (each counterclockwise step is chosen arbitrarily to show a basic free energy drop). The first-order rate constants of any model of this type imply a set of basic free energy levels, as in Eqs. (9.18) and Fig. 3.1(a). Equation (9.19) imposes a restraint on the rate constants chosen for the model since $\mu_S - \mu_P$ is generally a known quantity.

The set of basic free energy levels 1231 is repeated indefinitely above and below those shown in the figure, one set for each cycle.

Figure 3.1(b) shows an equilibrium case. Here we start with Fig. 3.1(a), hold c_P and μ_P constant, and decrease c_S and μ_S until $\mu_S = \mu_P$. The result is a single set of three equilibrium basic free energy levels. Correspondingly, $p_1^e > p_3^e > p_2^e$ (Boltzmann distribution).

Using this same model, we now consider an explicit example in which we discuss free energy dissipation and compare basic and

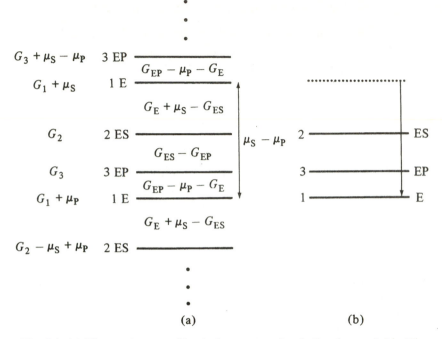

Fig. 3.1. (a) Illustrative set of basic free energy levels for the model in Fig. 1.4(b). $G_3 = G_{EP}$, etc. The levels are repeated above and below those shown. (b) An equilibrium set of basic free energy levels, derived from Fig. 3.1(a) (c_P constant; c_S decreased).

gross free energy levels (this requires some numerical calculations— to obtain the p_i).

Figure 3.2(a) shows the hypothetical set of basic free energy levels we use, with nonhorizontal lines indicating possible transitions. The location of the zero of free energy is arbitrary. Note that, in this example, $\Delta G'_{ij}$ is not positive in every counterclockwise step. In a particular model, any choice of α's must satisfy Eq. (9.19). Figure 3.2(b) gives a corresponding set of steady-state gross free energy levels (calculated as described below). As illustrated in Fig. 3.2(b), the total gross free energy drop for one circuit, the

98

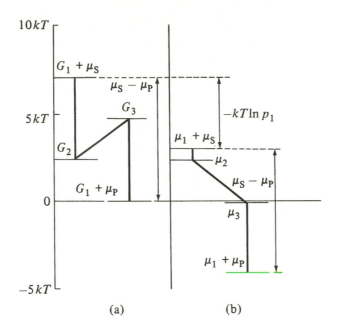

Fig. 3.2. (a) Basic free energy levels in a numerical example based on Fig. 3.3. (b) Gross free energy levels for the same case.

sum of the $\Delta\mu'_{ij}$, is also $\mu_S - \mu_P$. Both sets (basic and gross) of levels are repeated indefinitely, above and below, but this is not shown.

The above property of the gross free energies (i.e., sum $= \mu_S - \mu_P$) obtains even in a transient (i.e., at arbitrary t), as shown in Eq. (9.14). It should be pointed out, however, that in transients $\Delta\mu'_{ij}$ is not necessarily positive for every counterclockwise step. For example, suppose that, at $t = 0$, $p_3 = 1$ and $p_1 = p_2 = 0$. Then, for $t > 0$ but small we would clearly have $J_{23} < 0$ and hence $\Delta\mu'_{23} < 0$.

We turn now to the steady-state situation. As there is only one cycle in the diagram, we must have $J_{12} = J_{23} = J_{31} \equiv J$ at steady state. Since

$$\Phi = \Sigma J_{ij}\Delta\mu'_{ij} = J\Sigma\Delta\mu'_{ij} = J(\mu_S - \mu_P) > 0 \qquad (9.20)$$

and $\mu_S - \mu_P > 0$, we also have $J > 0$ (i.e., the net flux is in the direction of the force). Further, since $J\Delta\mu'_{ij}$ $(ij = 12, 23, 31) > 0$, we deduce that $\Delta\mu'_{ij}$ $(ij = 12, 23, 31) > 0$. Thus, at *steady state*, the gross free energy level must decrease [Fig. 3.2(b)] and the net flux must be positive (and equal) for each step in the direction of the force $(S \rightarrow P)$. Further, because J is constant around the cycle, the relative contribution of each transition to the overall dissipation of free energy is proportional to $\Delta\mu'_{ij}$. The results in this paragraph obviously apply to *any single-cycle model*.

There is no fundamental complication when a single cycle contains more than one force (e.g., $\mu_S - \mu_P$ above plus the force $\mu_A - \mu_B$ from the concentration gradient of a ligand, where A and B refer to the two sides of a membrane). The *net* force determines the direction of positive flux.

The rate constants in Fig. 3.3 have been chosen to be consistent with Fig. 3.2(a). We have $K_{12} = 100$, $K_{23} = 0.1$, $K_{31} = 100$, and $\mu_S - \mu_P = kT \ln 1000$. Given the rate constants, we can easily calculate the steady-state probabilities [Eqs. (6.10)–(6.12)]: $p_1 = 0.01855$, $p_2 = 0.972$, and $p_3 = 0.00901$. Most of the enzyme accumulates in state 2 $(= ES)$ because of the relatively small rate constants for transitions out of state 2. From Eq. (9.15), the mean flux in each step of the cycle is $J = 0.0882\alpha$. The gross free energy levels [Fig. 3.2(b)] are obtained from the basic levels [Fig. 3.2(a)] by subtraction of $-kT \ln p_i$ in each case. The gross level drops in

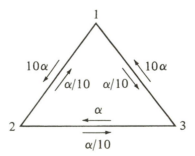

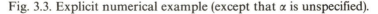

Fig. 3.3. Explicit numerical example (except that α is unspecified).

each counterclockwise step, as required. The rate of free energy dissipation, $J(\mu_S - \mu_P)$, is $0.609\alpha kT$.

10. Kinetic and Thermodynamic Formalism for Muscle Contraction[1]

In this section we give a brief account of the theoretical foundation of muscle contraction, assuming the correctness of the well-known sliding-filament model. The discussion will be based, for simplicity, on a three-state kinetic cycle, though a realistic diagram[5,6] would be more complicated than this. The primary objective here is to see, in a rigorous but not complicated way, how ATPase kinetics is linked to the mechanics of muscle contraction. In this free energy transduction system, some of the free energy of hydrolysis of ATP is converted into mechanical work.

This same kind of formalism can be applied to the cilia-flagella[7] (dynein) system and to fast axonal transport[8] (kinesin).

It is assumed that the reader is acquainted with structural and other general features of the sliding-filament model of muscle contraction.

We consider an ensemble of independent and equivalent cross-bridges in the overlap zone (Fig. 3.4), each of which has accessible to it at most one actin attachment site at a time. We use the term "cross-bridge" to apply to a projection from a myosin filament,

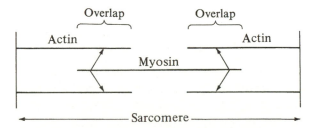

Fig. 3.4. Schematic sarcomere, showing myosin and actin filaments, overlap zones, and force exerted by cross-bridges on actin filaments.

101

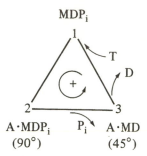

Fig. 3.5. Three-state ATPase attachment-detachment cycle used to illustrate the theoretical formalism for muscle contraction.

whether it is attached to an actin site or not. A cross-bridge can exist in several biochemical states, as illustrated in Fig. 3.5, where M is the myosin cross-bridge, A the actin site, T = ATP, and D = ADP. This particular kinetic diagram has only three states, but it is adequate to illustrate general principles. Two states are "attached" states (M attached to A) and one is an "unattached" state. One cycle in the positive direction hydrolyzes one molecule of ATP (T → D + P_i). The transition 3 → 1 encompasses several transient intermediates: this single transition includes, successively, release of D, binding of T, detachment of M from A, and hydrolysis of T. The three-state cycle involves not only hydrolysis of ATP but also, concomitantly, attachment and detachment of M to and from A: the transition 1 → 2 is an attachment transition whereas 2 → 3 is a detachment transition.

Unattached and Attached States

Figure 3.6(a) shows, schematically, adjacent myosin and actin filaments, and one cross-bridge protruding from the myosin filament. This cross-bridge is in state 1 and is not attached to an actin site. It has a Gibbs free energy G_1 (Section 9).

In Fig. 3.6(b), the cross-bridge is attached to an actin site at x, where x locates the position of the site *relative to the cross-bridge*,

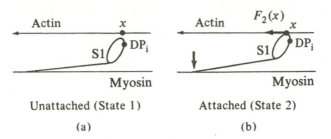

Fig. 3.6. (a) Illustration of state 1 (unattached). (b) Illustration of state 2 (attached at x).

using some convenient origin, for example, the short arrow at the bottom of the figure. By convention: x increases to the right; in contraction, the actin filament moves to the left (top left arrow) relative to the myosin filament. The cross-bridge, in this figure, is in state 2, with Gibbs free energy $G_2(x)$, a function of x. The variable x is essential in this problem; it is the feature that distinguishes muscle contraction from the systems treated in previous sections.

Each M contains two S1 fragments, or "heads." We assume, for simplicity, that only one head can be active at a time and hence we treat M formally as if it had only one head.

Force Exerted in an Attached State

In Figure 3.6(b), the top right arrow indicates the force $F_2(x)$ exerted by the cross-bridge on the actin filament when it is attached, in state 2, at x. If this were state 3, the force would be $F_3(x)$. No force can be exerted by M on A in state 1 (unattached). For any attached state i ($i = 2, 3$ in this example),

$$dG_i = F_i dx \quad (T, p \text{ constant})$$

$$\text{(10.1)}$$

$$F_i(x) = (\partial G_i / \partial x)_{T,p}.$$

Thus, $G_i(x)$ determines $F_i(x)$. We are assuming explicitly here that the cross-bridge, when in state i, is in internal equilibrium and can be treated as a small thermodynamic system. We shall use subscript i for an attached state and subscript j (or k) for *any* state.

Assumptions of the Sliding-Filament Formalism

With the above introduction, we list here some of the assumptions used in the formalism to follow.

(a) At any instant a given cross-bridge has accessible to it, for attachment with significant probability, only a *single* actin site. (b) The cross-bridge behaves operationally as if it has only one head. (c) Cross-bridges in the overlap zone act independently of each other. (d) A cross-bridge can exist in several different discrete biochemical states (providing the diagram), some attached to actin and some unattached, and transitions between these states include the binding and splitting of ATP.

Implicit in (d) are the following: the states of the diagram are in general not in equilibrium *with each other*, yet each state is an equilibrium state *internally*; transitions between states occur instanteously (on the time scale of the diagram); and force exerted by a cross-bridge on an actin filament is associated with *attached states* and *not with transitions*.

Basic Free Energy Changes

States 2 and 3 are attached states; hence G_2 and G_3 are functions of x. However, G_1 is independent of x. In general, we would expect all of the rate constants α_{jk} in Fig. 3.5 to be functions of x, because any transition involves at least one attached state. (If states 1 and 2 were both unattached states, α_{12} and α_{21} would *not* depend on x.) Following Eqs. (9.4) and (9.7), we have, at each x (i.e., for each location of the actin site relative to the cross-bridge),

$$\frac{\alpha_{12}(x)}{\alpha_{21}(x)} = \exp\left[\frac{G_1 - G_2(x)}{kT}\right] = K_{12}(x)$$

$$\frac{\alpha_{23}(x)}{\alpha_{32}(x)} = \exp\left\{\frac{G_2(x) - [G_3(x) + \mu_P]}{kT}\right\} = K_{23}(x) \qquad (10.2)$$

$$\frac{\alpha_{31}(x)}{\alpha_{13}(x)} = \exp\left\{\frac{[G_3(x) + \mu_T] - (G_1 + \mu_D)}{kT}\right\} = K_{31}(x).$$

The chemical potentials μ_T, μ_D, and μ_P were introduced in Eqs. (2.21) and (2.22). The quantities in Eqs. (10.2) that depend on x should be noted. The free energy G_j ($j = 1, 2, 3$) refers here to a *single* independent cross-bridge (a "small" thermodynamic system) fixed in the myofilament structure. Thus G_j has nothing to do with the "concentration" of cross-bridges in this structure. Note also that the "ligand" actin is not represented in these equations by μ_M terms, as would be the case for a solution system. This is because the actin site is now part of the permanent myofilament structure and is not a ligand in solution with a concentration.

If we multiply Eqs. (10.2) together, around the cycle, we obtain

$$\frac{\Pi_+(x)}{\Pi_-(x)} = \exp\left(\frac{\mu_T - \mu_D - \mu_P}{kT}\right) = e^{X_T/kT}$$

$$= K_{12}(x)K_{23}(x)K_{31}(x), \tag{10.3}$$

where X_T is a constant, independent of x.

In model building,[5] one usually begins with the free energy functions (of x). The rate constant functions and force functions (for attached states) must then be chosen to be consistent with the free energies, as prescribed by Eqs. (10.1) and (10.2).

Figure 3.7 is an explicit example of free energy functions, taken from reference 5. The actual functions are

$$G_1/KT = 20.00, \quad (G_3 + \mu_P)/kT = x^2/2\sigma^2$$

$$G_2/kT = 16.00 + [(x - 80)^2/2\sigma^2] \tag{10.4}$$

$$\sigma^2 = 200 \text{ Å}^2, \quad X_T/kT = 23.00,$$

where x is in Å. Note, from Eq. (10.1), that F_2 and F_3 are linear in x.

105

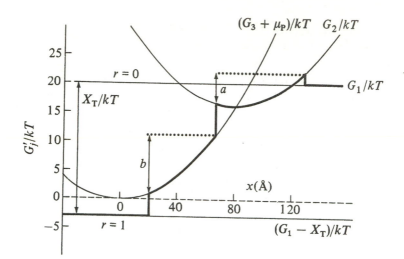

Fig. 3.7. Set of basic free energy levels based on Eqs. (10.4). The zero of free energy is arbitrary. These levels are repeated above and below, though not shown. See the text for details.

The model on which these equations are based is the following. The attached cross-bridges can exist in two different conformational states, referred to as the 90° (state 2) and 45° (state 3) states. In the 90° conformation, the cross-bridge exerts zero force (minimum in G_2) when it is attached to the actin site at a 90° angle. Elasticity resides in this angle: any other angle is less stable and has a higher free energy. Similarly, the 45° conformation exerts zero force (minimum in $G_3 + \mu_P$) when attached at 45°. These properties are illustrated in Fig. 3.8.

The top three levels in Fig. 3.7 may be extended above and below indefinitely (just as in Section 9). However, only one repeat (state 1) is included in the figure; this shows the value of X_T/kT.

Stochastics, Work, and Efficiency

Suppose we label the top three levels in Fig. 3.7 as the $r = 0$ set, the next set below [beginning with $(G_1 - X_T)/kT$] as the $r = +1$ set,

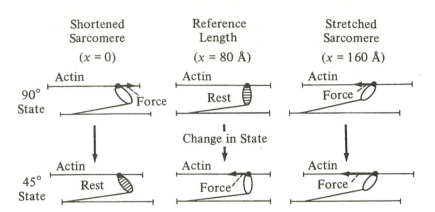

Fig. 3.8. Schematic representation of the attached cross-bridge. The 90° and 45° states are shown at three different values of x, the axial position of the actin site to which the cross-bridge is attached. The magnitude and direction of the force exerted by the cross-bridge is indicated by the length and direction of the horizontal arrows along the actin filament. The cross-bridge is shaded at its minimum free energy for that state (exerts no force).

etc. Thus we have the possible sets $r = 0, \pm 1, \pm 2, \ldots$. In an *isometric contraction*, if we consider just those cross-bridges with some particular x value, any one of these cross-bridges will perform a biased random walk on the basic free energy levels at x. The transition probabilities in the walk are the $\alpha_{jk}(x)$. The general trend of the walk will be downhill, but uphill transitions will sometimes occur. If a system (cross-bridge) starts, say, in state 1 in set $r = 0$ and ends in state 1 at some $r > 0$, after many transitions, r cycles will have been completed and r molecules of ATP (in solution) will have been hydrolyzed to products. The associated free energy loss is rX_T. Moreover, under isometric conditions (v, the velocity of contraction, is zero), no work is performed by any of the cross-bridges (though they exert force when in attached states).

In a *steady isotonic contraction*, one can imagine a single actin site appearing on the right side of Fig. 3.7 and moving to the left side of this figure, at velocity v, past a single cross-bridge. During the "pass" of the actin site by the cross-bridge (which is in the initial state 1), the cross-bridge may interact with the site via the transitions of the diagram and thus progress through some sequence of states. The transition probabilities $\alpha_{jk}(x)$ are continually changing during the pass (because x changes), as are the associated basic free energy levels (Fig. 3.7). The number of cycles completed during the pass may be $r = 0, \pm 1, \pm 2, \ldots$, with the most common result $r = 1, 2, \ldots$ (this depends on the velocity v).

The heavy path in Fig. 3.7 shows an example of a possible stochastic sequence of events in a case with $r = 1$. Transitions (instantaneous) are represented by vertical steps, up or down. Work is done by the cross-bridge on the actin filament only when the cross-bridge is in an *attached* state. The amount of work in the example in Fig. 3.7 is $a + b$, these being the net basic free energy decreases of the cross-bridge while in the attached states, 2 and 3, respectively. The relation between work and free energy follows from (for state i)

$$\text{work in} - dx = F_i(-dx) = F_i v dt = -dG'_i, \tag{10.5}$$

in view of Eq. (10.1) and $v = -dx/dt$.

As far as this one pass, with $r = 1$, is concerned (Fig. 3.7), the efficiency is clearly $(a + b)/X_T =$ work done/free energy loss. The transduction of free energy (at the molecular level) is the result of the cyclic biochemical behavior of the cross-bridge, being driven by the ATP free energy drop and accomplishing work by attachment to actin. The whole cycle is involved in the transduction process, and acts as an indivisible unit. The analogy to active transport of a ligand by ATP, through the mediation of an enzyme complex, is obvious.

Clearly, in order for the overall efficiency to be large, in most passes much of the total free energy drop rX_T must be accounted for by the cross-bridge riding the basic free energy levels of attached states downward. That is, discontinuous free energy drops associated with transitions, and upward rides, must be avoided as much as possible.

The efficiency is the stochastically averaged fraction of the total free energy drop that is due to net downward rides on the free energy levels of attached states. More precisely, it is the averaged work divided by the averaged free energy drop [compare Eq. (10.15)]. This is not as simple a recipe for calculating the efficiency as $\bar{F}v/\bar{J}X_T$ in Eq. (10.15) below, but it does provide an intuitive and very useful way of visualizing the efficiency.

Ensembles and Subensembles

We consider an ensemble (a very large number) of cross-bridges in the overlap zone. These can be divided into subensembles that contain cross-bridges with the actin site at the same x (more precisely, with the site between x and $x + dx$). Because of the lack of register between the cross-bridges on a myosin filament and the actin monomers of an actin filament, the distribution in the x values of the actin sites "seen" by the cross-bridges of the ensemble is uniform, or constant. Thus, all subensembles contain the same number of cross-bridges (if dx is chosen the same for each subensemble). Furthermore, this number is time invariant, irrespective of the type of experiment being considered, for if the sites are moving relative to the cross-bridges, as many cross-bridges enter the subensemble $x, x + dx$ as leave it.

The actin sites we are considering here have some repeat distance d. Since, by assumption (a) above, the cross-bridge can interact with only one site, d is about $7 \times 55 = 385\text{Å}$ (the distance along the actin filament between actin monomers in equivalent orientations). We can regard x as locating the position of the actin site "nearest" to a given cross-bridge, with $x = 0$ the "central" position. Then there are other non-nearest actin sites at $x + d, x - d$, etc. The nearest site is necessarily in the range $-d/2 \le x \le +d/2$, with uniform probability of occurrence at any x in this interval. Thus, in averaging over x values below, this is the appropriate interval to use, with uniform weighting in x.

Actually, an implication of the single-site assumption is that there is zero probability of cross-bridge attachment to the actin site when it is at the ends of the interval ($x = \pm d/2$). If necessary, the origin $x = 0$ should be shifted to assure this. For otherwise, the cross-bridge might be attached at either $x = +d/2$ or $-d/2$, contrary to our one-site assumption.

It should perhaps be emphasized that the single-site assumption is not likely to be realistic, but it is used for simplicity and because all the essential features of the problem appear even in this simplified model.

Differential Equations in State Probabilities

Consider the subensemble of cross-bridges with nearest actin site between x and $x + dx$, at time t. Of these cross-bridges, let $p_j(t, x)$ be the fraction in state j at t, where $j = 1, 2, 3$ in Fig. 3.5. Of course, $\Sigma_j p_j = 1$ since a cross-bridge has to be in some one state at any time t.

The time evolution of the subensemble at x is governed by a set of three first-order differential equations in the p_j, only two of which are independent (because of $\Sigma_j p_j = 1$). For example, for state 1,

$$\frac{dp_1}{dt} = \alpha_{21}(x)p_2 + \alpha_{31}(x)p_3 - [\alpha_{12}(x) + \alpha_{13}(x)]p_1, \quad (10.6)$$

etc. We shall not write down the companions of Eq. (10.6) because the whole set of differential equations follows automatically from the diagram.

In Eq. (10.6), dp_1 is the change in $p_1(t, x)$ owing to transitions that occur in the infinitesimal interval dt. But, if the sites happen to be moving relative to the cross-bridges at the rate dx/dt at t (all sites necessarily move together), $p_1 + dp_1$ is the value of p_1 at $t + dt$ *and at* $x + dx$, i.e., $p_1(t + dt, x + dx)$. This provides another expression for dp_1, and Eq. (10.6) becomes

$$\left(\frac{\partial p_1}{\partial t}\right)_x + \left(\frac{\partial p_1}{\partial x}\right)_t \frac{dx}{dt} = \alpha_{21}(x)p_2 + \alpha_{31}(x)p_3 - [\alpha_{12}(x) + \alpha_{13}(x)]p_1,$$

$$(10.7)$$

etc. The rate $dx/dt \ (\equiv -v)$ is the same for all j and all x, but it may be a function of t (in some transients). The left-hand side of Eq. (10.7) simplifies in special cases, as we shall see below.

110

Given the rate constants and the boundary conditions (the latter are partially determined by the nature of the experiment), we can, in principle at least, solve the differential equations to obtain $p_j(t, x)$ for all j, all $t \geq 0$, and $-d/2 \leq x \leq d/2$.

In one-actin-site models, the boundary conditions would include $p_i = 0$ (attached states) at $x = \pm d/2$, attachment rate constants $= 0$ at $x = \pm d/2$, detachment rate constants $= \infty$ at $x = \pm d/2$, and $p_j(d/2) = p_j(-d/2)$ (unattached states), because of the actin site periodicity.

The rate constants must be furnished as part of whatever molecular model is being used. Only the rate constants for transitions between unattached states are directly related to and potentially available from solution biochemistry. The other rate constants depend not only on the molecular architecture of the filament structure but also on the particular value of x (i.e., on the position of the actin site relative to the cross-bridge).

A most important point in this connection is that the rate constants are not all independent of each other. In fact each pair α_{jk} and α_{kj} is related via equations like (10.2) to the basic free energy levels (which generally have been preassigned). Further, Eq. (10.3) must be satisfied at each x. Finally, since the force functions $F_i(x)$ for attached states are also connected to the basic free energies $G_i'(x)$ by $F_i = \partial G_i'/\partial x$, care must be taken—in model building—to make the rate constants and force functions consistent with each other.

Force Exerted on the Actin Filament

When a cross-bridge of the subensemble at x is attached in state i to the actin site, the force exerted on the actin filament by the cross-bridge is $F_i(x)$, as given by Eq. (10.1). This is a consequence of the fact that state i is in internal equilibrium. $F_i(x)$ is the force exerted by the cross-bridge regardless of the nature of the experiment in progress (isotonic contraction, etc.). It is a strictly molecular property, not dependent on macroscopic external constraints such as load, etc.

The mean force (per cross-bridge) exerted on the actin filament, at t, by the cross-bridges in the subensemble at x, is then

$$F(t, x) = \sum_i p_i(t, x)F_i(x), \qquad (10.8)$$

where the $p_i(t, x)$ are solutions of the differential equations above. That is, there is a contribution to this mean force from each *attached* state i.

The mean force (per cross-bridge) exerted on the actin filament at t by *all* cross-bridges in the ensemble is then found by averaging $F(t, x)$ over x:

$$\bar{F}(t) = \frac{1}{d} \int_{-d/2}^{+d/2} F(t, x)dx. \qquad (10.9)$$

Steady Isometric Contraction

We consider here the steady-state force exerted by the cross-bridges in the overlap zone on the actin filaments when the length of the muscle fiber is held constant. Then in the set of Eqs. (10.7), we put $(\partial p_j/\partial t)_x = 0$ for all j ("steady") and $dx/dt = 0$ ("isometric"). Thus, at each x, we have the linear algebraic equations

$$0 = \alpha_{21}(x)p_2 + \alpha_{31}(x)p_3 - [\alpha_{12}(x) + \alpha_{13}(x)]p_1, \qquad (10.10)$$

etc., which can be solved (together with $\Sigma_j p_j = 1$) to give each $p_j(x)$ as a function of all the rate constants. The diagram method of Chapter 2 may be useful. Substitution of the $p_i(x)$ and $F_i(x)$ in Eqs. (10.8) and (10.9) leads to $\bar{F}$, the mean steady isometric force generated per cross-bridge (in the overlap zone).

Steady Isotonic Contraction

In an experiment of this kind, one fixes the load and then observes the early steady velocity of contraction, $v = -dx/dt = $ const. The opposite point of view is more convenient in a theoretical calculation: the parameter v is specified in advance and the steady force generated per cross-bridge, $\bar{F}$, is then calculated from Eq. (10.9) (for the given v). This generated force is of course equal to the load (per cross-bridge in the overlap zone) that can be lifted by the muscle fiber contracting at velocity v.

In Eqs. (10.7) we put $(\partial p_j/\partial t)_x = 0$ ("steady") and $dx/dt = -v$. Since the p_j are functions of x only (and the parameter v), $(\partial p_j/\partial x)_t$ can be written dp_j/dx. Thus we have

$$-v\frac{dp_1}{dx} = \alpha_{21}(x)p_2 + \alpha_{31}(x)p_3 - [\alpha_{12}(x) + \alpha_{13}(x)]p_1, \quad (10.11)$$

etc. Steady isometric contraction is the special case $v = 0$. The solutions $p_i(x)$ of this set of equations, together with the force functions $F_i(x)$, then give $\bar{F}$ as a function of v from Eqs. (10.8) and (10.9). This is the so-called force-velocity relation (as calculated theoretically, from a model). The maximum velocity of contraction, v_{max}, is the value of v that gives $\bar{F} = 0$ (no load). This property is clearly independent of the *number* of cross-bridges acting, as is found experimentally.

Work and Efficiency in Steady Isotonic Contraction

In a steady isotonic contraction the force (and external load) per cross-bridge is $\bar{F}$ and the velocity of contraction is v. Since the work done (per cross-bridge in the overlap zone) in lifting the load, when the actin filaments move a distance $-dx$ relative to the myosin filaments, is $-\bar{F}\, dx$, the rate of performance of work, per cross-bridge, is simply $dW/dt = \bar{F}v$.

The free energy source for this work is the ATP in solution. We need the rate at which ATP is consumed, per cross-bridge in the overlap zone, by the ensemble of cross-bridges. Any one of the net transition fluxes in Fig. 3.5 can be used to calculate this rate (they all give the same result, *after* averaging over x). Let us use the transition $2 \rightarrow 3$, as an example (experimentally, this corresponds to measuring the rate of release of P_i to the solution). The rate of appearance of P_i, per cross-bridge in the subensemble at x, is

$$J(x) = \alpha_{23}(x)p_2(x) - \alpha_{32}(x)p_3(x), \quad (10.12)$$

where $p_2(x)$ and $p_3(x)$ have been found by solving Eqs. (10.11) at v. This transition flux at x has to be averaged over x to obtain the mean net flux per overlap cross-bridge:

$$\bar{J} = \frac{1}{d} \int_{-d/2}^{+d/2} J(x)dx. \tag{10.13}$$

The rate of loss of ATP free energy (per overlap cross-bridge) is then $\bar{J}X_T$ while the rate at which work is performed is $\bar{F}v$. Hence the rate of free energy dissipation (in ensemble plus bath plus load) is

$$\Phi = \bar{J}X_T - \bar{F}v \geq 0 \tag{10.14}$$

and the efficiency is

$$\eta = \frac{\bar{F}v}{\bar{J}X_T} = \frac{\bar{F}d}{\bar{r}X_T} \leq 1. \tag{10.15}$$

Of course $\bar{F}v = 0$ either in an isometric contraction ($v = 0$) or when $v = v_{max}$ ($\bar{F} = 0$). Some intermediate v is associated with the optimal efficiency. Experimentally, the optimal efficiency is of order 50%. The second expression for η in Eq. (10.15) follows from the stochastic point of view discussed above, where $\bar{r}$ is the mean number of cycles per pass: $\bar{F}d$ is the mean work done per pass and $\bar{r}X_T$ is the mean ATP free energy drop per pass. Thus, $\bar{r}$ and $\bar{J}$ are related by $\bar{r} = \bar{J}d/v$.

Reference 5 contains a detailed numerical calculation that illustrates the above formalism.

Discussion

It is worth emphasizing again that the force generated or exerted by cross-bridges on the actin filaments, and therefore the work done, is associated with biochemical *states* (specifically, attached states) and *not with transitions* or biochemical "steps." In general, each attached state will contribute to the force, though the relative magnitudes (and signs) of the contributions will depend on the probabilities of these states, the force functions, parameters such as the velocity of contraction v, and the value of x. The overall contribution of a given attached state to the force or work can be found only by averaging over x. Consideration of only one value of x will almost certainly be inadequate, even qualitatively. Averaging over x will reflect the fact that the relative importance of the different cycles and states of the diagram will ordinarily vary considerably with x.

Just as the work done by the cross-bridges will usually have to be attributed in some measure to all attached states and all values of x, the free energy provided by ATP splitting cannot be localized within the diagram or at any one x.

A cross-bridge with nearest actin site instantaneously at x has possible transitions and rate constants as specified by the diagram at x. But the cross-bridge has no way of "knowing" the kind of experiment being conducted. Put another way, the diagram at x is independent of the boundary (experimental) conditions that are to be used in solving the differential equations implied by the diagram. Specifically, in steady isotonic contraction, the rate constants in the diagram at any x do *not* depend on the load or force (per cross-bridge) $\bar{F}$ or on the corresponding velocity of contraction v.

References

1. Hill, T. L. (1977) *Free Energy Transduction in Biology* (Academic, New York).
2. Hill, T. L. and Eisenberg, E. (1981) Q. Revs. Biophys. **14**, 463.
3. Hill, T.L. (1983) Proc. Natl. Acad. Sci. USA **80**, 2922.
4. Hill, T. L. (1974) Prog. Biophys. Mol. Biol. **28**, 267.
5. Eisenberg, E., Hill, T. L. and Chen, Y. (1980) Biophys. J. **29**, 195.
6. Eisenberg, E. and Hill, T. L. (1985) Science **227**, 999.
7. Brokaw, C. J. (1976) Biophys. J. **16**, 1013.
8. Chen, Y. and Hill, T. L. (1988) Proc. Natl. Acad. Sci. USA **85**, 431.

Index

117

efficiency of, 108, 114
formalism for, 101
isometric, 107, 112
isotonic, 107, 112
mechanics of, 101
sliding-filament model for, 101, 104
velocity of, 107, 112
work in, 108, 113
Myosin, 101, 109
subfragment one, 103

P
Product, 9

R
Random walk, 7, 23, 64, 68, 71
Rate constants, 111
first-order, 6, 23, 90
pseudo first-order, 7, 91, 92
second-order, 7, 92
Reciprocal relations, 32

S
Slippage, 5, 11, 27, 29
State
attached, 101, 108

of complex, 4
probability, 8, 14, 20, 39, 43
unattached, 101
visits to, 65
Steady state, 1, 8, 14
Stochastics, 10, 62, 64, 68, 93, 108
Stoichiometry, 4, 5, 11, 19, 22, 28, 32, 33
Substrate, 9

T
Tennis, 84
Thermodynamics
irreversible, 29, 30
linear, 29, 36
non-equilibrium, 29
second law of, 24, 25, 36, 96
small system, 104, 105
Time average, 9
Tracers, 61, 80
Transient, 14, 99
Transitions, 4, 6
direction of, 95
isomeric, 90
psuedo-isomeric, 92
Transport, 4, 18
active, 4, 108